Immunoassay Technology
Volume 1

Immunoassay Technology

Volume 1

Editor S. B. Pal

Walter de Gruyter · Berlin · New York 1985

Editor

S. B. Pal, D. Phil., Dr. rer. biol. hum. M. I. Biol.
Universität Ulm
Department für Innere Medizin
Steinhövelstraße 9
D-7900 Ulm
F. R. of Germany

CIP-Kurztitelaufnahme der Deutschen Bibliothek

Immunoassay technology / ed. S. B. Pal. – Berlin ; New York : de Gruyter
NE: Pal, Skrikanthabhushen B. [Hrsg.]
Vol. 1 (1985).
ISBN 3-11-010062-2 (Berlin ...)
ISBN 0-89925-082-3 (New York)

ISBN 3 11 010062 2 Walter de Gruyter · Berlin · New York
ISBN 0-89925-082-3 Walter de Gruyter, Inc., New York

 Printing: Gerike GmbH, Berlin. – Binding: Dieter Mikolai, Berlin. – Cover-Design: Hans Bernd Lindemann, Berlin. – Printed in Germany.

Preface

Much of the existing evidence from published literature on *immunoassay* without the introduction of radioisotope has shown that this has now become an *established technology.* Therefore, it is felt that the time has come for *publication of a regular series of review volumes on nonisotopic immunoassay techniques,* which also prevent radioactive hazards in laboratories. It is hoped that this series will be of use to those who are engaged in this line of work and will also be of help in avoiding time-consuming literature survey.

I take this opportunity to thank Dr.-Ing. Rudolf Weber of Walter de Gruyter & Co., Berlin · New York, for agreeing to publish this series. I also thank Mrs. M. R. Pal for her assistance as an honorary editorial secretary.

March 1985 S. B. Pal

Contents

RECENT DEVELOPMENTS IN MEASURING URINARY CONSTITUENTS BY NONISOTOPIC IMMUNOASSAY TECHNIQUES

R. Trinick and F. Laker
Department of Clinical Biochemistry, Royal Victoria Infirmary, Newcastle upon Tyne, NE1 4LP, U.K.

Introduction

Urine estimation has formed part of the practice of medicine since earliest times. Comprehensive analysis of urine includes macroscopic and microscopic examination, microbiological and biochemical testing. Reasons for making accurate measurements of urinary constituents fall into two categories: a) to detect systemic disorders where abnormal substances or abnormal quantities of normal constituents are excreted in the presence of normal renal function, and b) to identify disease of the genito-urinary tract.

The ability to differentiate normal from minimally elevated levels of urinary constituents can be important clinically, for instance in detecting early diabetic nephropathy by measuring urinary excretion of small amounts of albumin (1). By identifying patients at risk and treating intensively, it may be possible to reverse the nephropathy (2). As many diabetic patients now die prematurely of renal failure this could be of immense clinical significance.

Until recently the measurement of low levels of urinary constituents has been limited by the sensitivity of the available analytical methods. However, many sensitive techniques have recently been described including radioimmunoassay (3) and non-isotopic immunoassay techniques, the latter group including heterogeneous (4, 5, 6, 7) and homogeneous (8) enzyme immuno-

Immunoassay Technology, Volume 1

assay and latex immunoassay (9). The purpose of this chapter is to review recent uses of nonisotopic immunoassay procedures for determining urinary composition. The general conditions for carrying out immunoassay procedures are well covered elsewhere (10, 11, 12) and will not be further considered here.

The advantages of nonisotopic immunoassay are well known and include: high sensitivity, high specificity, the use of cheap reagents with a long shelf life, inexpensive equipment, the variety of different enzyme labels which may be used allowing different conjugation techniques, no radiation hazard, homogeneous enzyme assays which can be automated and are rapid to perform.

Disadvantages include: constituents or preservatives in urine samples may affect enzyme activity, homogeneous enzyme assays are largely limited to measurement of haptens, and, heterogeneous enzyme assays require an antigen or antibody to be attached to a solid phase.

Several recent developments have led to improvements in nonisotopic immunoassay techniques, including higher quality plastic microtitration plates. These show less edge effects and have good adsorption properties. Semi-automated equipment is now available for dispensing reagents and for plate washing. Vertical light path photometers may be obtained which include dual wavelength measurement and enable compensations to be made for well to well variations. Such instruments may also include sophisticated data handling capabilities.

Urine Protein Measurement

Myoglobin

A competitive heterogeneous ELISA has been used by Cloonan et al. (13) to measure human myoglobin in both serum and urine. Rabbit anti-human myoglobin antiserum was used as the first antibody, goat anti-rabbit immunoglobulin serum as the second antibody, the enzyme being horseradish peroxidase. After incubating sample or standard with the first antibody, myoglobin-horseradish peroxidase conjugate was added. After further incubation the antibody-myoglobin-conjugate complexes were precipitated with the second antibody and the precipitate washed twice before adding substrate. The assay showed no cross-reactivity with other haem proteins and gave recoveries between 88% and 90%. Within assay precision was 7.4% and between assay precision was 11.2%. Sera stored at -20°C was stable for 24 weeks. However, myoglobin in urine stored at -20°C declined significantly over 24 weeks. The method gave similar results to a radioimmunoassay procedure. In patients with acute myocardial infarction, serum myoglobin was found to rise earlier than creatinine kinase and it was suggested that it may be valuable in the early diagnosis of acute myocardial infarction. Urinary myoglobin excretion was variable and no attempt was made to determine infarct size using myoglobin excretion with time urine sampling.

Ferritin

Ishikawa and colleagues (14) have developed a sandwich type enzyme immunoassay for ferritin determination in serum and urine. Urine collections were preserved with 0.1% sodium azide and stored at -20°C. Serum samples were stored at -20°C. The assay had a lower limit of sensitivity of 0.2 and 1 ng/ml in urine and in serum respectively. The within assay precision for serum was between 6% and 10% and the between assay precision was between 8% and 12%, at three levels. The within assay precision for urine was between 11% and 21% and the between assay precision was 10%-13% at three levels. Analytical recoveries for serum ferritin ranged from 95% to 100%, recoveries for urine ferritin ranged from 102% to 118%. A high correlation was found between levels of ferritin in urine and in serum and it was suggested that urine ferritin may provide a useful survey tool for evaluating iron stores. The method showed good correlation with serum ferritin determined by radioimmunoassay.

Retinol-binding protein

Retinol-binding protein (RBP) has been measured by a double-antibody sandwich type ELISA using an alkaline phosphatase conjugate (15). Urine was diluted 40-fold and serum 5000-fold to get it within the working range of the assay. The method correlated very well with rocket immunoelectrophoresis and radial immunodiffusion, however, recovery values were not quoted. The within assay precision was 5% at two levels, the between assay precision was 8% at two levels. Mean urine RBP concentration in a healthy group of 30 subjects was 101 µg/g creatinine (range 37.3 to 185 µg/g creatinine).

Lauwerys and colleagues (16) have described a latex immunoassay for RBP. Samples were incubated at 37°C for 30 minutes with antibody-coated latex particles and the resulting agglutination quantified by particle counting. The method also correlated well with rocket immunoelectrophoresis and radial immunodiffusion. Recoveries averaged 102% in urine and 93% in serum. Between and within assay coefficients of variation (CV) ranged from 5% to 11%. In healthy subjects the mean urinary excretion of RBP by this method was 52.5 µg/g of creatinine (SD = 59.2 µg/g of creatinine). The mean concentration in serum was 46 mg/l (SD = 10.4 mg/l).

It is suggested that RBP, like β2-microglobulin, may be a sensitive index for screening for tubular proteinuria (15). RBP circulates bound to prealbumin (17). However, after delivery of vitamin A to epithelial tissues RBP loses its affinity for prealbumin and is rapidly eliminated through the kidneys. In healthy subjects urinary excretion of RBP is small but is increased when renal tubular function is impaired (18). In a comparison of 68 patients with varying types of renal disease, levels of urinary RBP and β2-microglobulin showed a close correlation (19).

β2-Microglobulin

An immunoassay based on direct agglutination by β2-microglobulin (β2-m) of latex particles on which antibody against β2-m is adsorbed, has been described (20); β2-m was purified from the urine of the patient who developed a renal tubular lesion after exposure to cadmium. Antibody was raised in rabbits and coated onto latex particles which were stable for several months if stored at 4°C. The latex particles tend to aggregate spontaneously. However, this can be suppressed by disaggregating by vortex mixing just before the assay and stabilizing by adding bovine serum albumin. The pH of the assay solution and the concentration of free latex particles in the incubation mixture must be optimised for each batch of latex. The time required for optimal agglutination must be determined for each batch of latex particles.

Analysis was undertaken by incubating samples or standards with anti-β2-m coated latex particles and unagglutinated latex particles were counted. The number of unagglutinated particles was expressed as a percentage of the total number of free particles obtained with the zero standard. The lower limit of detection was 0.5 µg/l. Within assay precision in urine and serum at two different dilutions ranged from 4.6% to 8.7%. Between assay precision in urine and serum ranged from 8.4% to 10%.

Analytical recovery of β2-m in urine averaged 97% and in serum 104%. The assay has also been automated (21) and the method correlated well with radioimmunoassay. Mean concentration of β2-m in serum and urine from 33 healthy men averaged 1.6 mg/l and 54 µg/g of creatinine respectively; β2-m measurement in serum is a sensitive index of impaired glomerular filtration rate and in urine is an indication of tubular dysfunction.

Albumin

A modified double antibody sandwich ELISA technique carried out on microtiter plates has been described for measuring urinary albumin (22). The lower limit of sensitivity was 3.1 µg/l enabling albumin excretion in normal or pathological urine to be measured, using commercially available reagents. Between assay precision ranged from 8.8% to 9.2% and within assay precision was 4.0% at two levels. Analytical recovery ranged from 95% to 104%. Urine was collected without preservative and stored at -20°C until assayed after the addition of 20 µl of inactivated rabbit serum to prevent adsorption of albumin to the storage tubes. For twenty normal children the mean urinary albumin excretion was 7.6 mg/24 hr (range 1.7 - 22.9 mg/24 hr).

Bernard and Lauwerys (23) developed a latex immunoassay for urinary albumin whereby latex particles coated with rabbit anti-albumin immunoglobulin were incubated with the sample or standard at 37°C for 30 min and the resulting agglutination measured by particle counting. The lower limit of the standard curve was 25 µg/l. Between assay precision ranged from 8.2% to 11% and within assay precision ranged from 9% to 11.6% at three levels. Analytical recovery was 95% (range 80%-107%) and the method showed good correlation with nephelometry. Urine samples were stored at -18°C after addition of sodium azide. For 50 healthy adults the mean urinary albumin concentration was 6.82 mg/l (SD= 5.73) or 4.43 mg/g creatinine (SD = 2.98). The method is similar to that developed for β2-microglobulin (16) and also depends on the stability of the coated latex particles.

β-Thromboglobulin

β-Thromboglobulin (β-TG) is a low molecular weight protein originating from platelets and it has been suggested that elevated serum and urine levels may be an indicator of platelet activation (24) : urine minimises the possibility of artefactually releasing β-TG from platelets. A sandwich enzyme immunoassay for β-TG has been described (25) where β-TG was isolated from human platelets and anti-β-TG serum raised in rabbits. The purified antiserum was digested with pepsin to obtain the $F(ab')_2$ fragments which were immobilised non-covalently on polystyrene beads. The enzyme labelled specific antibody was prepared by conjugating $F(ab')_2$ fragments with β-D-galactosidase from Escherichia coli. After incubating the antibody coated beads with sample or standard, they were washed twice with buffer and incubated with the enzyme-labelled specific antibody. Following a further washing step, a fluorogenic substrate was added and the resulting fluorescence measured. twenty-four hour urine collections were made without any preservative. The minimum detectable amount of β-TG was 10 pg/ assay tube. The results of this assay correlated well with those of a radioimmunoassay. Plasma within run precision ranged from 4.7% to 9.7%, between assay precision ranged from 6.3% to 8.4%. Urine within assay precision ranged from 7.4% to 14.1%, between run precision ranged from 6.3% to 11.6%. Plasma β-TG levels in healthy subjects had a mean of 38.4 ng/ml (SD = 12.2). Urinary β-TG levels were 0.17 ng/ml (SD = 0.05).

Immunoglobulins

Double antibody sandwich ELISA techniques to measure human IgG, A and M in urine have been described by Trinick and Laker (26) using commercially available material. Urine was collected overnight without any preservative and stored at -20°C or 4°C if analysed the next day. Urine immunoglobulin concentrations and excretion rates were measured in 14 healthy men and are shown in Table I.

TABLE I

Urine immunoglobulin concentrations and excretion rates in 14 healthy adult men

	IgG		IgA		IgM	
	mg/l	mg/mmol creatinine	mg/l	mg/mmol creatinine	mg/l	mg/mmol creatinine
Mean	0.28	0.03	0.24	0.02	0.05	0.005
Range	0.11–0.48	0.008–0.06	0.03–0.57	0.006–0.06	0 –0.40	0 –0.03

The mean value of urinary creatinine was 11.6 mmol/l, range 4.7–20.5 mmol/l.

Urine immunoglobulin concentrations and excretion rates in six patients with gross proteinuria are shown in Table II.

TABLE II

Urine immunoglobulin concentrations and excretion rates in six patients with gross proteinuria

	IgG		IgA		IgM	
	mg/l	mg/mmol creatinine	mg/l	mg/mmol creatinine	mg/l	mg/mmol creatinine
Mean	450	63	112	14	1.5	0.2
Range	54–740	2–160	14–250	0.5–30	0.01–3.7	0.0005–0.47

The mean value of urinary creatinine was 12.7 mmol/l, range 5.3–25.8 mmol/l. The mean value of urinary protein was 6.5 g/l, range 4.7–9.8 g/l.

Mean urinary immunoglobulin excretion rates in 14 healthy men were: IgG 19 ± 2.7 µg/h, IgA 15 ± 2.4 µg/h, IgM 4 + 2.2 µg/h (mean ± SEM). Within assay precision was IgG 4.6%, IgA 8.1%, IgM 7.0%. Between assay precision was IgG 4.2%, IgA 4.5%, IgM 9.4%. Analytical recoveries were mean ± range, IgG 95% (78%-118%), IgA 101% (96%-112%), IgM 106% (93%-120%).

Figure 1 shows a standard curve obtained for IgG. Curves for IgA and IgM were similar. The standards used were commercially available protein standards. The lower limit of sensitivity for these assays were IgG 20 µg/l, IgA 10 µg/l, IgM 25 µg/l.

Singh and Makker (27) have developed an ELISA for immunoglobulin G (IgG) in body fluids. Protein A containing Staphylococcus aureus has affinity for IgG of most mammalian species. To measure IgG in CSF or urine, 100 µl of Staphylococcus aureus suspension, 100 µl of 1:800 diluted goat anti-human IgG labelled with alkaline phosphatase were added sequentially to 100 µl of diluted urine or standard and incubated for 30 minutes. After centrifuging, decanting and washing the bacterial pellets, p-nitrophenyl phosphate substrate was added and the absorbance read at 405 nm. The method gave good correlation with radial immunodiffusion although the Staphylococcus

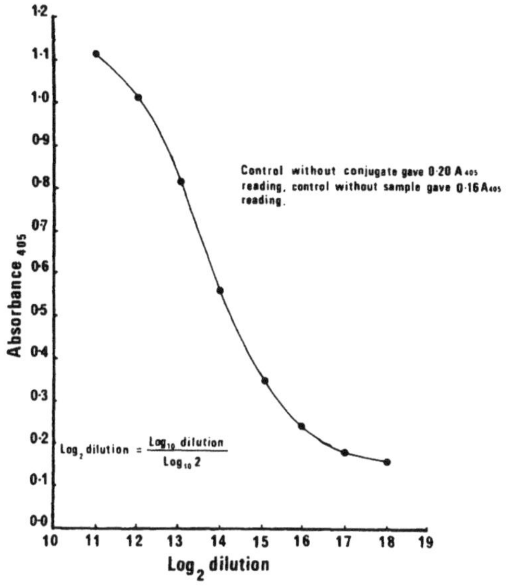

Fig. 1. A standard curve for IgG (curves for IgA and IgM are similar).

aureus ELISA had much greater sensitivity. Between assay precision was 16.8%-20.7% and within assay precision was 4.4%-5.7% at two levels. The lower limit of sensitivity was 10 µg/ml. It was pointed out that protein A-containing Staphylococcus aureus binds human IgG_1, IgG_2 and IgG_4 but not IgG_3. In addition, Fab, Fab_2 or light chains inhibit the binding of anti-IgG-enzyme conjugate to the bacterial-bound IgG, a problem common to other immunological methods.

Hormones in Urine

Total oestrogens

Messeri et al. (28) have described a homogeneous luminescence immunoassay for "total" oestrogens in enzymatically hydrolysed urine from non pregnant women. Hydrolysis was accomplished by overnight incubation of urine at 37°C with "Helicase" (beta-glucuronidase and sulphatase) or 1 hour incubation at 50°C with a higher concentration of enzyme. The sample or standard was incubated for 1 hour at room temperature with the chemiluminescent tracer (oestradiol-17-hemisuccinate-aminobutylethyl isoluminol) and antiserum which had been raised against oestriol 16, 17-dihemisuccinate-BSA conjugate. (The antiserum specifically bound to C-19 steroids carrying the phenolic group characteristic of oestrogens). The sample was then placed in a luminometer with microperoxidase and hydrogen peroxide added to initiate light emission. The standard used was 17β-oestradiol. The lower limit of sensitivity in the assay was 4 µg/l of 17β-oestradiol. The mean analytical recoveries at three levels of 17β-oestradiol were 88.1%, 95.7% and 92.3%. The within assay precision was 6.5%-6.7%, the between assay precision 8.6%-9.8%. The method correlated well with a conventional fluorimetric method and showed very little cross-reactivity with urinary steroids other than oestrogens. Fresh early morning urine collected without any preservative was used and patients could collect and store samples at home, before bringing them to the clinic. After incubating for 1 hour, daily monitoring of ovulation was possible.

Luteinizing hormone

A double antibody sandwich technique for measuring luteinizing hormone (LH) using antisera to LH conjugated with alkaline phosphatase has been described by Rathnam et al. (29). Purified anti-LH was absorbed onto 1/4 inch polystyrene beads or onto polystyrene tubes, the latter being easier to handle. The assay was conducted by simultaneously adding to the coated bead or tube, sample/standard and antibody enzyme conjugate, or adding sample/standard first, incubating, then adding the antibody-enzyme conjugate. The simultaneous protocol gave a faster assay with greater colour development. Urine was diluted 5-fold to lessen the effect of interfering substances. It is not clear whether random urine samples or 24 hr urine samples were used.

Analytical recoveries of known amounts of LH varied from 88% to 92%. Four replicate determinations gave C.V. values between 95% and 105% of the mean LH value. The authors compared their method with a radioimmunoassay method and found that the ELISA gave higher values during the LH surge than the radioimmunoassay. The assay takes 90 minutes to perform and provides a simple method for detecting and quantifying the LH surge.

Drug Measurements

Dapsone

Huikeshoven et al. (30) have described an ELISA inhibition method for detecting dapsone in urine. The assay was developed to monitor the compliance of patients in taking their medication. This is a sensitive method which can be used in leprosy-endemic countries where laboratory services are often limited. The wells of a standard microtiter plate were coated with a diazotised dapsone/horseshoe-crab haemocyanin conjugate. Urine sample was added, double diluted down the plate and, after incubation, anti-dapsone rabbit serum was added. The plates were then washed and horseradish-peroxidase-conjugated anti-rabbit IgG antiserum added. Those urines containing dapsone failed to develop colour with substrate addition. The assay was more sensitive than conventional spectrophotometric procedures. Urine was preserved with hydrochloric acid.

Cannabinoids

A homogeneous enzyme immunoassay for measurement of cannabinoid metabolites and Δ^9-tetrahydrocannabinol (Δ^9-THC) in urine has been developed (31). The antibody reagent was an antisera, raised in sheep, to an aldehyde derivative of Δ^9-THC conjugated with bovine serum albumin. The enzyme reagent was Δ^9-THC conjugated with malate dehydrogenase from pig heart mitochondria.

Urine calibrator for the assay was Δ^9-THC diluted in a synthetic urine (a negative urine pool spiked with Δ^9-THC gave higher enzyme activity after lyophilization and storage, so a synthetic urine which did not show this effect was used). The assay detected several different cannabinoids but did not cross-react with various natural hormones, drugs or their metabolites. As the assay was used to detect cannabinoids in urine rather than measuring levels of cannabinoids in intoxicated individuals, a cut-off value of 15 µ/l was adopted as this level gave less than a 5% false positive rate. The within assay coefficient of variation ranged from 11.5% to 16.8%. The assay was rapid and easy to use. The authors found that urinary cannabinoids remained detectable for more than 24 hours after exposure, by this method.

New Techniques

Automation of latex immunoassay by particle counting

Bernard and Lauwerys (21) have described a continuous flow system to measure several proteins in either plasma or urine. The proteins measured were human ferritin, β_2 microglobulin, retinol-binding protein and albumin. Latex immunoassay is based on the agglutination of calibrated latex particles on which a specific antibody has been adsorbed by the protein against which the antibody was raised. The authors quantified the assay by counting the remaining unagglutinated particles (particle counting immunoassay) (9). The coated latex were stable at 4°C for over a year.

The samples were diluted and placed in assay cups on a sample tray. The sampler aspirated simultaneously the sample and the latex particles, using a peristalic pump. After mixing and incubating in a heated coil the stream was passed through the optical cell of the particle counter and the result displayed on a chart recorder. The peak height of each analysis was expressed as a percentage of the mean peak height of the zero standards. The total throughput time for each sample was 40 minutes. The lower limit of detection was 10^{-10} to 10^{-12} mol/l. Within and between assay C.V.'s were less than 10%.

The specificity and accuracy of this assay for β2-microglobulin, retinol-binding protein and albumin were similar to those already mentioned (16, 20, 23). Ferritin in serum at lower dilutions showed interference resulting in an underestimation. The authors recommended assaying at two different dilutions to circumvent this and to overcome the possibility of a post-zone effect in the albumin assay. Mean analytical recovery for ferritin was 104%.

Conclusions

The specificity and sensitivity of immunological methods means that useful and reliable information can be obtained on a urine sample. The ready and increasing commercial availability of non isotopic reagents is an important step. Their stability and long shelf life, coupled with easily accomplished assay procedures which are becoming steadily more automated, means that these techniques are beginning to be easier, cheaper and more reliable.

In the clinical laboratory, blood is preferred to urine, as it is a more direct sample, and urine collection can be unreliable. However, there are several situations in which a urine sample may be more informative. In research on renal pathophysiology, urine is indispensable, and also in determining the extent of renal damage in general systems diseases such as the connective tissue diseases and diabetes mellitus (1). Urine analysis can be used to detect renal damage at an early stage in industries using nephrotoxic agents, for instance β2-microglobulin levels in cadmium workers (20). One of the early uses of non isotopic immunoassay procedures was to detect opiates in urine. Recently, cannabinoid detection in urine has become feasible, with the advantages of easy sample collection and detection some time after exposure (31).

The ease with which a urine sample can be collected could be an advantage. If many samples are required over several days it may be easier for patients to collect and store their own samples rather than repeatedly travelling to hospital for monitoring. This is of particular value in determining ovarian function with urinary oestrogens and luteinizing hormone (28, 29).

Similarly, in population surveys, it may be easier to obtain urine samples. Some techniques allow "do it yourself" kits to be produced with which sophisticated measurements can be made without expensive laboratory equipment. Urinary oestrogens could be estimated in a consulting room (28) or drug compliance assessed,without sophisticated laboratory facilities, in underdeveloped countries (30). Urine testing to monitor drug compliance could be greatly expanded with non isotopic immunoassay techniques.

These techniques allow more sensitive and specific urinalysis which will not only advance understanding of renal function in health and disease, but also allow better patient monitoring. No doubt these techniques will be applied to urine measurement in many other ways in future.

Acknowledgements

We gratefully acknowledge permission from Elsevier Biomedical Press B.V. to reproduce Tables I and II and Figure 1 (Clin. Chim. Acta 139, 113-117 (1984). Dr. Trinick was seconded from the Eastern Area Health and Social Services Board, Northern Ireland.

References

1. Viberti, G.C., Jarrett, R.J., Mahmud, U., Hill, R.D.: Microalbuminuria as a predictor of clinical nephropathy in insulin dependent diabetes mellitus. Lancet i, 1430-1432 (1982).

2. Viberti, G.C., Pickup, J.C., Jarrett, R.J., Keen, H.: Effect of control of blood glucose on urinary excretion of albumin and β_2-microglobulin in insulin-dependent diabetes. New Engl. J. Med. 300, 638-641 (1979).

3. Yalow, R.S., Berson, S.A.: Assay of plasma insulin in human subjects by immunological methods. Nature 184, 1648-1649 (1959).

4. Avrameas, S., Guilbert, B.: A method for quantitative determination of cellular immunoglobulins by enzyme-labelled antibodies. Europ. J. Immunol. 1, 394-396 (1971).

5. Engvall, E., Perlman, P.: Enzyme-linked immunosorbent assay (ELISA). Quantitative assay of immunoglobulin G. Immunochemistry 8, 871-874 (1971).

6. Engval, E., Perlman, P.: Enzyme-linked immunosorbent assay. III. Quantitation of specific antibodies by enzyme-labelled anti-immunoglobulin in antigen-coated tubes. J. Immunol. 109, 129-136 (1972).

7. Van Weeman, B.K., Schuurs, A.H.W.M.: Immunoassay using hapten-enzyme conjugates. FEBS letters 24, 77-81 (1971).

8. Rubinstein, K.E., Schneider, R.S., Ullman, E.T.: "Homogeneous" enzyme immunoassay : a new immunochemical technique. Biochem. biophys. Res. Commun. 47, 846-850 (1972).

9. Cambiaso, C.L., Leak, A.E., De Steenwinkel, F., Billen, J., Masson, P.L.: Particle counting unimmunoassay (PACIA). I. A general method for the determination of antibodies, antigens and haptens. J. Immunol. methods 18, 33-44 (1977).

10. Schall, R.F., Tenoso, H.J.: Alternative to radioimmunoassay: Labels and methods. Clin. Chem. 27, 1157-1164 (1981).

11. Sullivan, M.J., Bridges, J.W., Marks, V.: Enzyme immunoassay : a review. Ann. Clin. Biochem. 16, 221-240 (1979).

12. Avrameas, S.: Enzyme immunoassays and related techniques : Development and limitations. Curr. Top. Microbiol. Immunol. 104, 93-99 (1983).

13. Cloonan, M.J., Bishop, G.A., Wilton-Smith, P.D., Carter, I.W., Allan, R.M., Wilcken, D.E.L.: An enzyme immunoassay for myoglobin in human serum and urine. Method development, normal values and application to acute myocardial infarction. Pathology 11, 689-699 (1979).

14. Ishikawa, K., Narita, O., Saito, H., Kato, K.: Determination of ferritin in urine and in serum of normal adults with a sensitive enzyme immunoassay. Clin. chim. Acta 123, 73-81 (1982).

15. Lucertini, S., Valcavi, P., Mutti, A., Franchini, I.: Enzyme-linked immunosorbent assay of retinol-binding protein in serum and urine. Clin. Chem. 30, 149-151 (1984).

16. Bernard, A.M., Moreau, D., Lauwerys, R.R.: Latex immunoassay of retinol-binding protein. Clin. Chem. 28, 1167-1171 (1982).

17. Peterson, P.A.: Demonstration in serum of two physiological forms of the human retinol binding protein. Europ. J. clin. Invest. 1, 437-444 (1971).

18. Peterson, P.A., Berggard, I.: Isolation and properties of a human retinol-transporting protein. J. biol. Chem. 246, 25-33 (1971).

19. Bernard, A.M., Lauwerys, R.R.: Retinol-binding protein in urine : A more practical index than urinary β_2-microglobulin for the routine screening of renal tubular function. Clin. Chem. 27, 1781-1782 (1981).

20. Bernard, A.M., Vyskocil, A., Lauwerys, R.R.: Determination of β_2-microglobulin in human urine and serum by latex immunoassay. Clin. Chem. 27, 832-837 (1981).

21. Bernard, A.M., Lauwerys, R.R.: Continuous-flow system for automation of latex unimmunoassay by particle counting. Clin. Chem. 29, 1007-1011 (1983).

22. Fielding, B.A., Price, D.A., Houlton, C.A.: Enzyme immunoassay for urinary albumin. Clin. Chem. 29, 355-357 (1983).

23. Bernard, A.M., Lauwerys, R.R.: Latex immunoassay of urinary albumin. J. Clin. Chem. Clin. Biochem. 21, 25-30 (1983).

24. Kaplan, K.L., Nossel, H.L., Drillings, M., Lesznik, G.: Radioimmunoassay of platelet factor 4 and β-thromboglobulin: development and application to studies of platelet release in relation to fibrinopeptide A generation. Br. J. Haematol. 39, 129-135 (1978).

25. Tanaka, M., Kato, K.: A sensitive enzyme immunoassay system for the measurement of β-thromboglobulin in plasma and urine. Clin. chim. Acta 123, 111-119 (1982).

26. Trinick, T.R., Laker, M.F.: Measurement of urinary immunoglobulins G, A and M by an enzyme-linked immunosorbent assay (ELISA). Clin. chim. Acta 139, 113-117 (1984).

27. Singh, A.K., Makker, S.P.: A rapid tube ELISA for human IgG in body fluids using Staphylococcus aureus. Clin. chim. Acta 129, 271-277 (1983).

28. Messeri, G., Caldini, A.L., Bolelli, G.F., Pazzagli, M., Tommasi, A., Vannucchi, P.L., Serio. M.: Homogeneous

luminescence immunoassay for total oestrogens in urine. Clin. Chem. 30, 653-657 (1984).

29. Rathnam, P., Saxena, B.B.: A "sandwich" solid-phase enzyme immunoassay for lutropin in urine. Clin. Chem. 30, 665-671 (1984).

30. Huikeshoven, H., Landheer, J.E.V., Denderen, A.C., Vlasman, H., Leiker, D.L., Das, P.K., Goldring, O.L., Pondman, K.W.: Demonstration of dapsone in urine and serum by ELISA inhibition. Letter in Lancet, i, 280-281 (1978).

31. Rodgers, R., Crowl, C.P., Eimstad, W.M., Hu, M.W., Kam, J.K., Ronald, R.C., Rowley, G.L., Ullman, E.F.: Homogeneous enzyme immunoassay for cannabinoids in urine. Clin. Chem. 24, 95-100 (1978).

ENZYME IMMUNOASSAY FOR DETERMINATION OF PANCREATIC GLUCAGON IN PLASMA

S. Iwasa
Chemistry Laboratory. Central Research Division, Takeda Chemical Industries Ltd., Osaka 532, Japan

Introduction

Glucagon is a 29-amino acid polypeptide hormone produced in the α cells of the pancreatic islets (1). The synthesis of glucagon is thought to involve a larger precursor, which is then enzymatically cleaved to the functional form, as with other polypeptide hormones. Molecules of various sizes displaying glucagon-like immunoreactivity have been found in gastric and intestinal mucosa (2, 3) and in plasma (4). Glucagon is primarily operative in the regulation of the plasma glucose concentration and also plays an important role in protein and fat metabolism, exerting a catabolic action in each case. Together with its major antagonist, insulin, glucagon has a central role to play in mediating nutrient balance (5). Such physiological studies became possible by the development of specific radioimmunoassay (RIA).

Many reports have appeared on radioimmunological determination in plasma (6-10) since Unger pioneered the RIA of glucagon (11). Initially, values of 300-400 pg/ml were recorded in fasting healthy subjects, but subsequently normal values of 50-150 pg/ml were reported. The difference between these two values was due to an immunologically cross-reacting protein called gut glucagon-like immunoreactivity (GLI), which reacted with certain glucagon antisera. GLI has none of the biologic actions of glucagon except for stimulating insulin release. RIA techniques using specific antibody for pancreatic glucagon

Immunoassay Technology, Volume 1

(PG) have been currently employed for determining plasma levels of glucagon.

However, no enzyme immunoassay (EIA) for glucagon had been reported in the clinical field until we developed a sensitive and specific EIA employing the different C-terminal peptides of glucagon for the preparation of the enzyme conjugate and the immunogen (12). PG-specific antibodies were prepared by immunizing rabbits with the C-terminal peptide (15-29) (13), and the enzyme conjugate was prepared by coupling the C-terminal peptide (21-29) to β-D-galactosidase. This EIA method was applied to determine glucagon levels in human and rat plasma, and was compared with an established RIA technique (14, 15).

The present work reviews the development of a sensitive and specific EIA for glucagon and its clinical application.

Materials and Methods

1. Glucagon and its related peptides

Seven different peptides were used for the preparation of enzyme conjugates and immunogens, and for the characterization of antisera.

Peptide I: pancreatic glucagon (1-29)
Peptide II: synthetic C-terminal glucagon fragment (15-29)
Peptide III: synthetic C-terminal glucagon fragment (15-29) containing norleucine instead of methionine at the third position from the C-terminus
Peptide IV: synthetic C-terminal glucagon fragment (21-29)
Peptide V: synthetic C-terminal glucagon fragment (22-29)
Peptide VI: synthetic C-terminal glucagon fragment (23-29)
Peptide VII: synthetic N-terminal glucagon fragment (1-10)

Purified porcine glucagon (peptide I) was purchased from Sigma Chemical Co., St. Louis, Mo. Glucagon fragment peptides II to VII were prepared by the conventional solution method (16). The analytical data for the products are shown in Table I

TABLE I. Analytical data for glucagon fragment peptides.

Pancreatic glucagon (peptide I, 1-29)

His^{1}-Ser-Gln-Gly-Thr^{5}-Phe-Thr-Ser-Asp-Tyr^{10}-Ser-Lys-Tyr-Leu-Asp^{15}-Ser-Arg-Arg-Ala-Gln^{20}-Asp-Phe-Val-Gln-Trp^{25}-Leu-Met-Asn-Thr^{29}

Peptide II (15-29):	$[\alpha]_D^{24}$=-30.0^{o} (c=0.3 in 50% AcOH); R_f 0.59[a] amino acid analysis: Arg 2.15; Trp 0.91; Asp 3.13; Thr 0.99; Ser 0.87; Glu 2.20; Ala 1.05; Val 0.96; Met 1.00; Leu 1.07; Phe 1.07 (79%)[b]
Peptide III (15-29):	$[\alpha]_D^{22}$=-30.9^{o} (c=0.3 in 50% AcOH); R_f 0.60 amino acid analysis: Arg 2.11; Trp 0.97; Asp 3.40; Thr 0.93; Ser 0.83; Gln 2.10; Ala 1.00; Val 1.04; Leu 0.91; Phe 0.84; Nle 1.06 (74%)
Peptide IV (21-29):	$[\alpha]_D^{23}$=-26.0^{o} (c=0.2 in 80% AcOH); R_f 0.76 amino acid analysis: Trp 1.07; Asp 2.06; Thr 0.98; Glu 1.06; Val 0.97; Met 1.00; Leu 1.04; Phe 1.11 (73%)
Peptide V (22-29):	$[\alpha]_D^{21}$=-23.5^{o} (c=0.5 in 80% AcOH); R_f 0.83 amino acid analysis: Trp 0.93; Asp 1.09; Thr 0.94; Glu 1.00; Val 0.90; Met 1.01; Leu 1.00; Phe 0.96 (81%)
Peptide VI (23-29):	$[\alpha]_D^{21}$=-15.9^{o} (c=0.5 in 80% AcOH); R_f 0.70 amino acid analysis: Trp 1.04; Asp 1.00; Thr 0.91; Glu 0.99; Val 1.07; Met 0.99; Leu 1.00 (78%)
Peptide VII (1-10):	$[\alpha]_D^{27}$=-35.0^{o} (c=0.5 in 1% AcOH); R_f 0.54 amino acid analysis: His 0.92; Asp 0.93; Thr 2.00; Ser 1.80; Glu 1.04; Gly 1.00; Tyr 0.96; Phe 0.99 (84%)

[a] Thin layer (cellulose) chromatography with n-butanol:pyridine:acetic acid:water (15:10:3:12 by volume). [b] Average recovery.

together with their structures. Gut GLI was extracted from canine jejunum according to Kenny's method (17)

2. Antisera

Antisera N2C and N6E were prepared by immunizing rabbits with C-terminal peptide II conjugated to bovine serum albumin (BSA) by means of glutaraldehyde (18). Antiserum N14B was prepared by using N-terminal peptide VII instead of peptide II. These peptide-BSA conjugates were emulsified with an equal volume of Freund's complete adjuvant and then injected at multiple subcutaneous sites along the dorsal surface. Three antisera 30K, G21 and R517 raised against PG were used for comparison. Although antisera 30K and G21 are specific for PG (19, 20), antiserum R517 cross-reacts with gut GLI (21).

3. Peptide-enzyme conjugates

PG and its fragment peptides were labelled with β-D-galactosidase (Boehringer Mannheim, West Germany) using metamaleimidobenzoyl N-hydroxysuccinimide ester (MBS) as a coupling agent (22). Briefly, PG or the fragment peptide was mixed with MBS in tetrahydrofuran, then the MBS-acylated peptide obtained was incubated with the enzyme. Next, the reaction mixture was directly chromatographed on a Sepharose 6B column using 0.02M phosphate buffer (pH 7.0) containing 0.1M NaCl, 1mM $MgCl_2$, 0.1% BSA, and 0.1% NaN_3 (buffer A) as an eluent. The fractions of eluate containing the peak of enzyme activity were used for glucagon assay.

They were stored in buffer A containing 1000 U/ml of aprotinin (Antagosan®, Hoechst, West Germany) with a less than 10% decrease in enzymatic activity and 10%-30% decrease in immunological activity within 5 months (14). The slight decrease had little influence upon the sharpness of glucagon standard curves. Moreover, a recent investigation has shown that the peptide-enzyme conjugates were stable for more than two years by lyophilizing *in vacuo* in the presence of sucrose or mannitol (unpublished data).

4. EIA Procedures

EIAs of PG were carried out by the double-antibody solid phase method using anti-rabbit immunoglobulin G-coupled cellulose particles (23). As an assay buffer, 0.02M phosphate saline (pH 7.3) containing 0.5% human serum albumin (HSA), 0.5% EDTA, and 0.1% NaN_3 (buffer B) was used. The incubation mixture was prepared by consecutive additions of 200 μl assay buffer, 50 μl (500 U) of aprotinin, 50 μl of either plasma sample or a buffered standard, 100 µl of antiserum and 100 µl of peptide-enzyme conjugate, giving a total volume of 500 µl in a 12 x 75 mm glass disposable tube (Corning Glass Works, Corning, N.Y. USA). The antiserum was diluted to the extent that it would bind approximately 30% of the enzyme activity. The incubation mixture was allowed to stand at 4°C for 4 days and 100 µl of 5% (W/V) suspension of an anti-rabbit immunoglobulin G-coupled cellulose were added to the solution. The suspension was then shaken at 30°C for 4 hr, followed by two washings with buffer A. The solid phase was resuspended in 500 µl of buffer A solution containing 10 µg of 4-methylumbelliferyl-β-D-galactoside (P-L Biochemicals Inc., Milwaukee, Wis. USA) as substrate (24) and incubated overnight at room temperature. The reaction was terminated by adding 3 ml of 0.1M carbonate buffer (pH 10.5) and the suspension was centrifuged. The fluorescence intensity of the supernatant was measured using a fluorimeter with excitation at 365 nm and emission at 450 nm.

5. RIA Procedures

Glucagon RIA was performed using dextran-coated charcoal and glucagon-^{125}I (Diabetes Research Fund, South Western Medical School and Nuclear Medical Laboratories, Dallas, Texas, USA) (25). As an assay buffer, 0.2M glycine buffer (pH 8.8) containing 0.2% HSA and 1% normal sheep serum was used. The total volume of the assay mixture was 1.2 ml, including 20 pg of labelled glucagon, 500 U aprotinin, and 0.2 ml of plasma sample.

RESULTS

1. Specificity of antisera against glucagon fragment

The specificities of antisera N2C and N6E raised against a C-terminal glucagon fragment (peptide II) were studied in EIA using peptide IV-enzyme conjugate. Figure 1 shows the reactivity of N6E with various C-terminal peptide fragments and gut

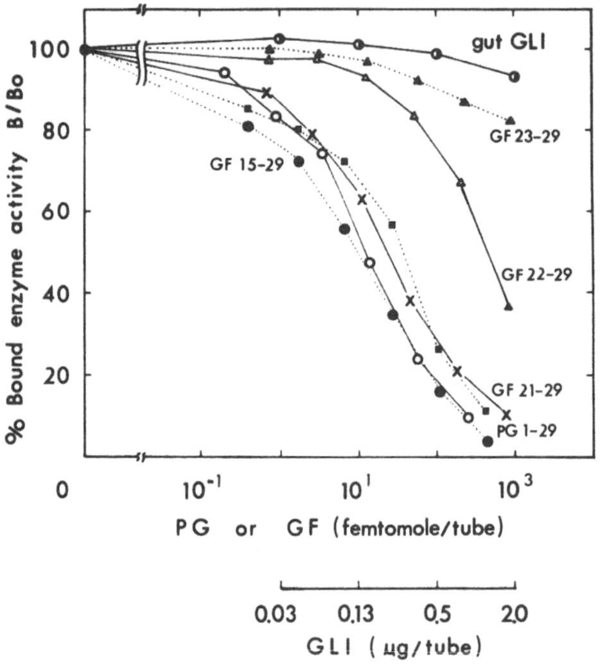

Fig. 1. Inhibition-binding curves of the enzyme-labelled peptide IV with standard pancreatic glucagon (PG), C-terminal glucagon fragments (GF; described in Table I) and gut glucagon-like immunoreactivity (GLI) in the enzyme immunoassay using antiserum N6E. PG (1-29, peptide I) (o), GF (15-29, peptide II) (●), GF (15-29, peptide III) (■), GF (21-29 peptide IV) (x), GF (22-29, peptide V) (Δ), GF (23-29, peptide VI) (▲), and gut GLI (◑). B_0 : Enzyme activity bound to the solid phase without added PG, GF or gut GLI. B: Enzyme activity bound to the solid phase with added PG, GF or gut GLI.

GLI extracted from canine jejunum. This antiserum bound to peptides I, II, III and IV much more strongly than to peptide V and VI. It did not cross-react with gut GLI and it was thus considered to be as PG-specific as Unger's 30K antiserum. Antiserum N2C gave similar results, although they are not shown in Fig. 1.

Thus, the antisera obtained by immunizing peptide II were specific for PG, as reported (13, 26). The probable reason is that the N-terminal and middle portions of the PG molecule contain the antigenic sites shared by gut GLI, as suggested (27, 28, 29). In fact, antiserum N14B raised against an N-terminal glucagon fragment (peptide VII) reacted with gut GLI as strongly as cross-reacting antiserum R517 raised against the whole molecule of PG.

2. Effect of combinations of antiserum and enzyme-labelled peptide on the sensitivity of PG-EIA

Van Weemen and Schuurs reported that the sensitivity of estrogen EIA could be increased when the estrogen derivatives used to produce the enzyme conjugate and the immunogen differed slightly (30, 31). We also investigated the effect of the combinations of antiserum and enzyme-labelled peptide in PG-EIA.

Peptides I to IV were each used as a peptide in the enzyme conjugate. Table II shows the effect of each combination of antiserum and enzyme-labelled peptide on the sensitivity of PG-EIA. Sensitivity is expressed as the midpoint, i.e. the concentration of PG which reduces the binding of the peptide-enzyme conjugate to antibody to half the value obtained in the absence of free PG. Antisera 30K and R517 gave acceptable sensitivity in two combinations, but the use of the antisera with the peptide II-enzyme conjugate resulted in about six times higher sensitivity than the conventional use of the antisera with peptide I-enzyme conjugate, despite the fact that more antibody was required to obtain the same degree of binding of the peptide-enzyme conjugate. Antisera 30K and R517 had such weak immune reactivity against peptide III- or IV-enzyme conjugate that the enzyme activity bound to the antisera was undetectable. The assays using N2C and N6E show that peptide I or IV is preferable as a peptide in the enzyme conjugate. In the N2C assay system, the peptide I- or IV-enzyme conjugate gave ten times more sensitivity than the peptide II-enzyme conjugate. Curves obtained

TABLE II. Midpoints for pancreatic glucagon (in ng/ml) in enzyme immunoassays using various combinations of antiserum and peptide-enzyme conjugates.

Peptide in immunogen[a]	Antiserum	Final dilution factor of antiserum	Peptide in enzyme conjugate[a]	Sensitivity midpoint[b] (ng/ml)
I	30K	40,000	I	1.1
		12,000	II	0.20
		1,600	III	-[c]
		1,600	IV	-[c]
	R517	32,000	I	1.8
		6,000	II	0.35
		1,600	III	-[c]
		1,600	IV	-[c]
II	N2C	60,000	I	0.88
		200,000	II	9.2
		60,000	III	2.9
		200,000	IV	0.85
	N6E	40,000	I	0.60
		200,000	II	2.2
		100,000	III	3.4
		200,000	IV	0.53

[a] Described in "MATERIALS AND METHODS." [b] Described in "RESULTS." [c] The binding of the peptide-enzyme conjugate to antibody was too weak for the detection of enzyme activity bound to the solid phase.

using 30K with peptide II-enzyme conjugate and N6E with peptide IV-enzyme conjugate were steeper than those obtained using 30K with peptide I-enzyme conjugate and N6E with peptide II-enzyme conjugate, particularly in the low concentration region of PG, where steeper curves result in increased sensitivity and precision (Fig. 2). The detection limit of the assays using 30K with the peptide II-enzyme conjugate and N6E with the peptide IV-enzyme conjugate was 20 pg/ml and 40 pg/ml, respectively.

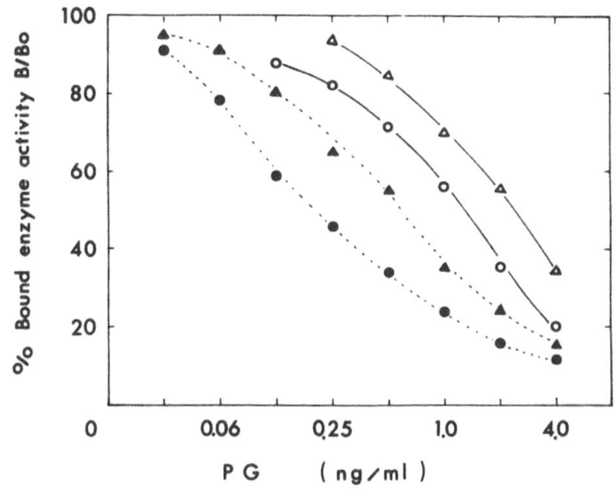

Fig. 2. Comparison of various combinations of antiserum and peptide-enzyme conjugates. In the 30K assay system, peptide I (o) or II (●) was used as a peptide in the enzyme conjugate, and in the N6E assay system, peptide II (Δ) or IV (▲) was used. Peptides I, II and IV are described in Table I. B_o and B are described in Fig. 1.

3. Precision and reproducibility of N6E-EIA

The EIA, using a combination of antiserum N6E and peptide IV-enzyme conjugate, was found to be as sensitive and specific as the conventional 30K-RIA (25). The precision and reproducibility of this N6E-EIA were further investigated.

Plasma specimens from rats which received intravenous arginine infusion at a dose of 1g/kg body weight for 30 min were serially diluted 1:1 to 1:16 with buffer B and assayed for PG by N6E-EIA. As shown in Fig. 3, the curves for standard PG and for the serially diluted plasma were parallel in specimens containing different levels of PG.

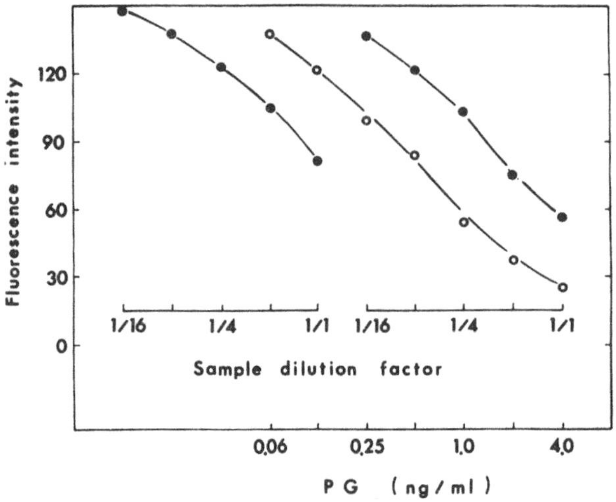

Fig. 3. Standard curve for pancreatic glucagon (PG) (o) and curves for serially diluted plasma specimens (●) measured by the enzyme immunoassay using antiserum N6E and peptide IV-enzyme conjugate (described in "Results").

Various amounts of PG were added to human sera to give final concentrations of 63-1000 pg/ml and the mixtures were assayed for PG by N6E-EIA (Table III). The recovery of PG added ranged from 76% to 146% with mean $\pm$ SD of 104%$\pm$ 17% for 15 determinations.

The coefficients of variation were measured at six different PG levels. They were 3.7%-14.5% (within assay) and 9.0%-18.5% (between assay). These results show that the N6E-EIA system is as reproducible and precise as the conventional RIA.

4. Comparison of EIA and RIA for measurement of glucagon in plasma

The immunoreactive glucagon (IRG) levels in 25 plasma samples were determined by both immunoassays of N6E-EIA and 30K-RIA

TABLE III. Recovery of pancreatic glucagon added to human sera.[a]

Serum no.	Added (pg/ml)	Estimated[b] (pg/ml)	Recovered (pg/ml)	Recovered (%)
1	-	62	-	-
	63	120	58	92
	125	201	139	111
	250	330	268	107
	500	570	508	102
	1,000	1,280	1,218	122
2	-	95	-	-
	63	168	73	116
	125	251	156	125
	250	330	235	94
	500	570	475	95
	1,000	1,100	1,005	101
3	-	102	-	-
	63	194	92	146
	125	232	130	104
	250	292	190	76
	500	590	488	98
	1,000	870	768	77

[a] Mean$\pm$S.D.=104$\pm$17%. [b] Each sample was assayed for glucagon by EIA using antiserum N6E at a final dilution of 1:200,000 with peptide IV-enzyme conjugate. Assay procedures were as described in "MATERIALS AND METHODS."

using glucagon-^{125}I. In the N6E-EIA, as described above, peptide IV-enzyme conjugate was used as a labelled antigen since the combination gave a sensitive assay. There was a good correlation ($r = 0.974$) between the values for IRG determined by the two methods in 25 samples (Fig. 4), but 19 of the 25 specimens showed lower levels in the EIA than in the RIA and the regression line determined by the method of least-squares was $y = 1.056\ x - 0.0452$. The tendency to give a slightly smaller IRG

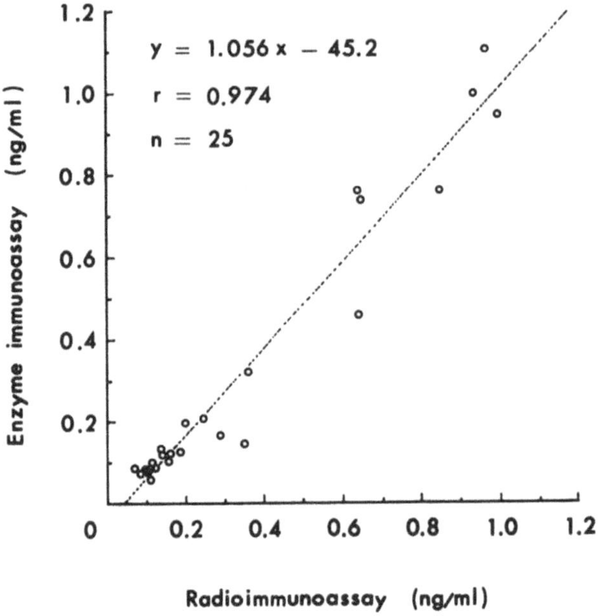

Fig. 4. Comparison of plasma glucagon levels obtained by radioimmunoassay (RIA) and by enzyme immunoassay (EIA). In the RIA antiserum 30K was used with glucagon-^{125}I, and in the EIA antiserum N6E was used with peptide IV-enzyme conjugate. The least-squares regression line has a slope of 1.056 and an intercept of -45.2 pg/ml. The correlation coefficient for these points is 0.974.

level in the N6E-EIA than in the 30K-RIA was remarkable at low levels in the basal state. The average values of the 14 basal samples measured by the N6E and 30K systems were 102 ± 16 and 135 ± 52 pg/ml, respectively.

5. Comparison of antisera N6E and 30K in EIA

The EIA system using N6E with peptide IV-enzyme conjugate was compared in plasma IRG levels with the conventional system

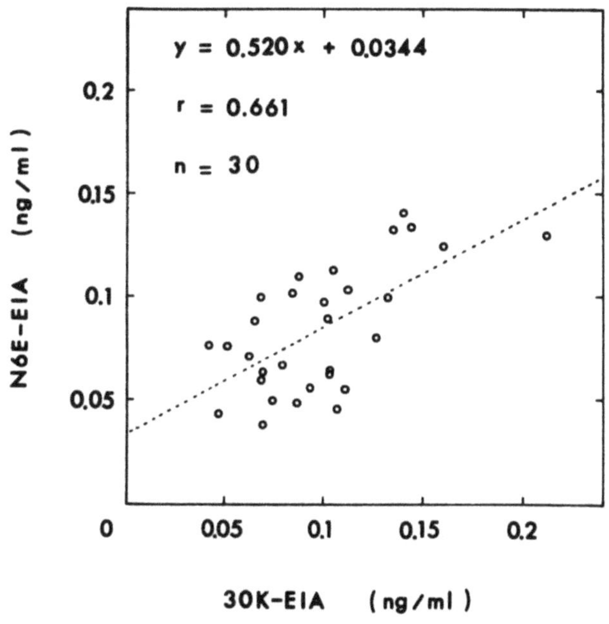

Fig. 5. Comparison of plasma glucagon levels obtained by enzyme immunoassays (EIA) using antisera 30K and N6E. In the 30K-EIA antiserum 30K was used with glucagon-enzyme conjugate and the N6E-EIA is described in Fig. 4. The least-squares regression line has a slope of 0.520 and an intercept of 34.4 pg/ml. The correlation coefficient for these points is 0.661.

using 30K with PG-enzyme conjugate (Fig. 5). Plasma specimens were obtained from human volunteers and rats in the basal state. The correlation (r = 0.661) obtained here was inferior to that obtained in the comparison between N6E-EIA and 30K-RIA, and the IRG levels in the N6E system (84 ± 23 pg/ml) were lower than those in the 30K system (98 ± 36 pg/ml). The latter results were similar to those obtained in the comparison of N6E-EIA and 30K-RIA.

For further investigations of the difference between N6E- and 30K-immunoreactivities, plasma specimens were applied to Bio Gel P-30 column chromatography using 0.02M phosphate saline (pH 7.3) containing 0.5% EDTA and 0.1% NaN_3 as eluent (4), and the eluates were assayed in both systems. Figure 6 shows the elution pattern of the IRG in the representative plasma with a high big plasma glucagon (BPG) level. Most of the IRG was recovered in the protein zone, which is the so-called "Big Plasma Glucagon", and only a small fraction appeared in the true glucagon area. It is noticeable that the BPG value in the 30K system was much higher than that in the N6E system. Table IV shows the IRG levels in the BPG and PG fractions from nine subjects, including normal human volunteers and arginine-infused rats. BPG and PG fractions obtained by gel filtration were pooled and concentrated by lyophilization followed by the two immunoassays. For BPG determination, the 30K system gave higher IRG values than the N6E system. For PG determination, however, the two systems gave similar values. These data mean that the use of N6E with peptide IV-enzyme conjugate resulted in less cross-reactivity of BPG than the use of 30K with PG-enzyme conjugate.

6. Application to glucose tolerance and arginine infusion tests

In order to investigate the applicability of the N6E-EIA system in the clinical field, glucose tolerance and arginine infusion tests were performed using plasma specimens from 18 normal human subjects and 13 patients with various diseases.

In the glucose tolerance test, after an overnight fast, 75 g of glucose was given orally to 12 normal subjects and 10 patients with diabetes mellitus, fatty liver, chronic hepatitis, obesity or polyp of the colon. Blood samples were obtained from the antecubital vein at 30 min intervals for 3 hr for the normal subjects and 2 hr for the patients (32). An arginine test was carried out in 6 normal subjects, 2 patients with diabetes mellitus and a patient with cirrhosis of the liver (33). After the collection of two base line samples, 10 min apart, 300 ml of

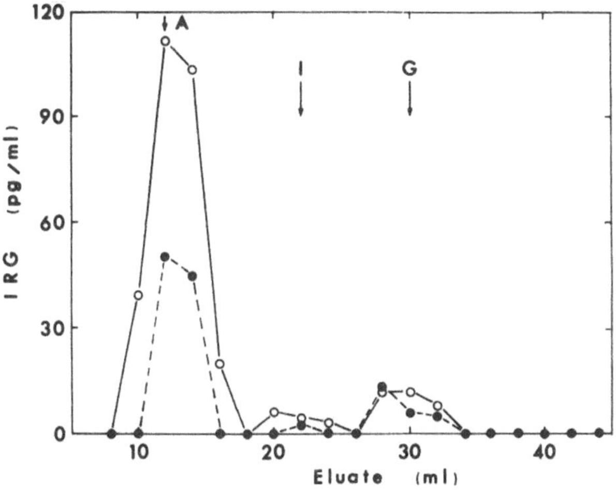

Fig. 6. Chromatographic pattern on a Bio Gel P-30 column of human plasma in the basal state. Immunoreactive glucagon (IRG) levels of the eluates were determined by enzyme immunoassays using antisera 30K (o) and N6E (●). Each assay system is described in Figs. 4 and 5. The arrows indicate the elution volume of the standard markers (A:bovine serum albumin, I:bovine insulin, G:porcine pancreatic glucagon).

10% arginine was infused for 30 min. Blood samples were drawn at 5, 10, 20, 30, 40, 50, 60, 75, 90 and 120 min after the arginine infusion. Blood samples were obtained from the antecubital vein with a heparinized syringe. Plasma glucagon (IRG) was measured both by immunoassay of N6E-EIA and RIA using PG-specific antiserum G21 raised against peptide I (20). Blood glucose and plasma insulin (IRI) were simultaneously measured by the glucose oxidase method (15) and the double antibody method of Morgan and Lazarow (34), respectively.

Figure 7 shows the changes in blood glucose, IRI and IRG during glucose tolerance tests in a group of 12 normal subjects. Blood glucose increased from the base line of 82 $\pm$ 1 mg/100 ml to a peak of 127 $\pm$ 3 mg/100 ml at 30 min and then decreased.

TABLE IV. Comparison of immunoreactive glucagon values[a] (pg/ml) for the big plasma glucagon (BPG) and pancreatic glucagon (PG) fractions in plasma measured by enzyme immunoassays using antisera 30K and N6E[b].

Plasma no.	BPG fraction		PG fraction	
	30K	N6E	30K	N6E
1	145	68	60	99
2	149	112	106	88
3	162	76	104	156
4	1,230	440	97	94
5	83	32	23	21
6	500	350	130	115
7	170	22	1,920	2,000
8	86	22	460	480
9	145	86	25	18

[a] They were expressed as the quantities to PG.

[b] 30K- and N6E-EIAs are described in Figs. 4 and 5.

Plasma IRI increased from the base line of 13 ± 2 μU/ml to a peak of 59 ± 8 μU/ml at 30 min and returned to the initial level by 180 min. Plasma IRG measured by G21-RIA fell from the initial level of 71 ± 7 pg/ml to a nadir of 53 ± 6 pg/ml at 60 min and returned to the base line at 180 min. Plasma IRG measured by N6E-EIA fell from the base line level of 100 ± 4 pg/ml to a nadir of 83 ± 8 pg/ml at 90 min and then returned to the initial level. There was a similar pattern of plasma IRG changes measured both by RIA and EIA. The results of patients are presented in Table V.

Changes in blood glucose, plasma IRI and plasma IRG during the arginine infusion test in a group of normal subjects are presented in Fig. 8. Blood glucose increased from the base line of 84 ± 3 mg/100 ml to 93 ± 5 mg/100 ml at 20 min and then fell

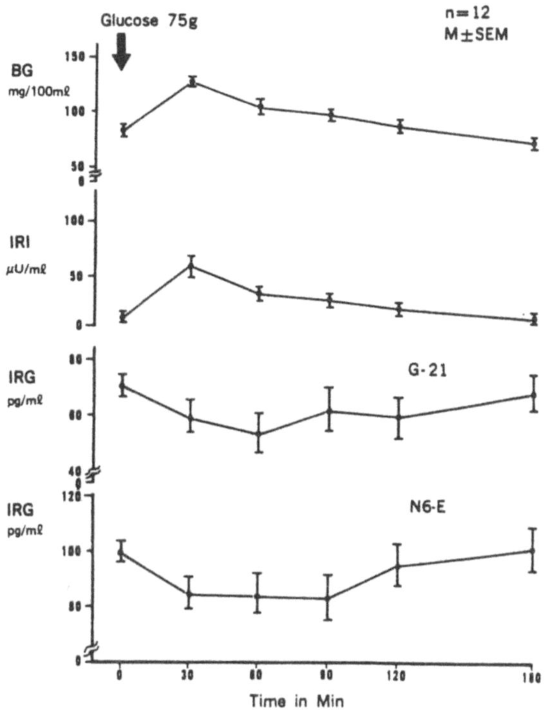

Fig. 7. Changes in blood glucose (BG), plasma insulin (IRI) and plasma glucagon (IRG) during glucose tolerance test in a group of 12 normal subjects. IRG was measured by radioimmunoassay using antiserum G21 and by enzyme immunoassay using antiserum N6E.

to 71 ± 5 mg/100 ml at 60 min. Plasma IRI increased from the base line of 26 ± 4 µU/ml to 99 ± 11 µU/ml at 30 min and returned to the initial level at 60 min. Plasma IRG measured by G21-RIA increased from the base line of 109 ± 7 pg/ml to 273 ± 59 pg/ml at 30 min and returned to the initial level at 120 min. Plasma IRG measured by N6E-EIA increased from the base line of 130 ± 13 pg/ml to a peak of 379 ± 60 pg/ml and fell to the initial level at 120 min. The changes in blood glucose and plasma

TABLE V. Comparison of plasma glucagon (pg/ml) measured by enzyme immunoassay (EIA)[a] and radioimmunoassay (RIA)[b] during glucose tolerance test in a group of 10 patients with various diseases.

Patients Diagnosis	Measure-ments[c]	Fasting	Minutes after glucose 30	60	90	120	Patients Diagnosis	Measure-ments[c]	Fasting	Minutes after glucose 30	60	90	120
U.Y.	BG	168	301	354	362	305	T.O.[d]	BG	89	195	167	153	133
Diabetes	RIA	85	75	50	25	80	Diabetes	RIA	70	50	53	45	50
mellitus	EIA	51	68	62	60	54	mellitus	EIA	55	40	40	41	62
S.T.	BG	142	233	311	357	372	A.K.	BG	69	131	152	178	187
Diabetes	RIA	80	80	50	45	30	Obesity	RIA	70	10	35	70	25
mellitus	EIA	77	89	58	45	59		EIA	94	65	59	60	70
S.Ss.	BG	88	166	270	291	320	K.H.	BG	70	120	117	76	72
Diabetes	RIA	95	60	65	45	40	Fatty	RIA	60	60	55	65	45
mellitus	EIA	87	54	50	47	43	liver	EIA	49	46	46	43	53
S.Se.	BG	108	195	228	225	218	K.I.	BG	103	141	203	136	126
Diabetes	RIA	90	60	60	105	60	Chronic	RIA	50	40	20	40	50
mellitus	EIA	113	95	99	96	100	hepatitis	EIA	46	34	34	37	43
H.W.	BG	165	238	368	368	342	T.M.	BG	128	144	180	169	154
Diabetes	RIA	55	80	50	45	80	Colon	RIA	70	30	10	20	15
mellitus	EIA	39	48	37	36	31	polyp	EIA	58	61	45	47	41

[a] In the EIA antiserum N6E was used with peptide IV (21-29)-enzyme conjugate. [b] In the RIA antiserum G21 was used with glucagon-^{125}I. [c] Venous plasma glucose, mg/100ml. [d] Glucose tolerance test was performed after treatment with sulfonylurea.

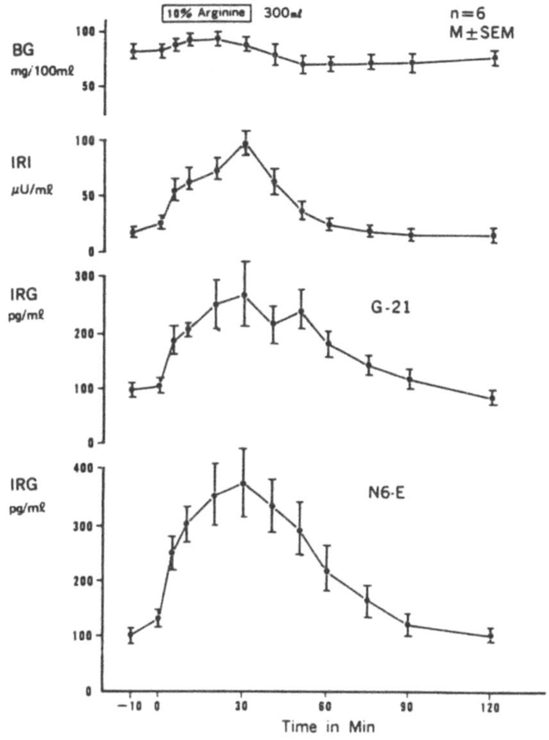

Fig. 8. Changes in blood glucose (BG), plasma insulin (IRI) and plasma glucagon (IRG) during arginine infusion test in a group of 6 normal subjects. IRG was measured by radioimmunoassay using antiserum G21 and by enzyme immunoassay using antiserum N6E.

IRG during the arginine test in 3 patients are presented in Table VI.

In order to investigate the validity of N6E-EIA, the correlation between plasma levels of IRG measured by these two assay methods was calculated. The correlation coefficient in the arginine test for the normal subjects was 0.93 and the regression equation y = 1.54 x - 43.5, while that in the group of patients was 0.98, with the regression equation y = 0.89 x - 31.3, where

TABLE VI. Comparison of plasma glucagon (pg/ml) measured by enzyme immunoassay (EIA)[a] and radioimmunoassay (RIA)[a] during arginine infusion test in three patients.

Patients	Measure-ments[b]	Minutes after arginine infusion											
		-10	0	5	10	20	30	40	50	60	75	90	120
K.K.	BG	215	208	211	216	221	219	229	253	269	267	251	236
Diabetes	RIA	65	60	155	305	400	445	325	-	120	145	100	100
mellitus	EIA	19	20	122	102	214	183	133	-	81	53	64	56
M.W.	BG	159	167	171	169	178	192	200	213	217	223	221	223
Diabetes	RIA[c]	10	10	70	65	50	70	50	65	35	10	10	10
mellitus	EIA	11	14	29	51	-	61	84	83	53	45	26	32
M.S.	BG	124	121	123	123	115	115	115	111	111	110	107	111
Liver	RIA	480	370	825	1650	1522	1400	1200	1375	1200	725	975	875
cirrhosis	EIA	369	306	891	1080	1340	1260	1030	1170	1150	580	871	823

[a] Described in TABLE V. [b] Venous plasma glucose, mg/100ml. [c] The value 10 represents the level less than the minimal detectable standard.

y was plasma IRG measured by EIA and x that measured by RIA.

Discussion

The development of a sensitive and specific EIA for glucagon and the applicability of EIA in the clinical field have been described. Various combinations of antibody and peptide-enzyme conjugates were used and it was found that the sensitivity could be increased by using peptide II-enzyme conjugate in an anti-PG system and peptide IV-enzyme conjugate in an anti-peptide II system (Table II, Fig. 2). These combinations improved the sensitivity of glucagon EIA to the extent that the technique is now comparable to RIA. The EIA was able to detect as little as one to two pg of glucagon in plasma. Such combinations may reduce the affinity of the antibody for the peptide-enzyme conjugate and thus give free glucagon at very low concentrations a better chance to compete.

The antisera obtained by immunizing peptide II were specific for PG, as reported (13), and they did not cross-react with gut GLI (Fig. 1). Such PG-specific antisera are useful in immunoassays for PG determination since plasma usually contains GLI of intestinal origin.

PG and the C-terminal peptide molecules were readily conjugated with β-D-galactosidase using MBS as a coupling agent. The enzyme was chosen as a good marker with a high specific activity and with functional thiol groups in the native form which can be coupled to peptides (35-37). Such enzyme-labelled peptides were stored at 4°C in buffer A with only a slight decrease in enzymatic and immunological activities, and the slight decrease hardly affected the sensitivity of EIA. The high stability of the enzyme-labelled peptides might result in good reproducibility and precision of the EIA method.

Plasma IRG levels of subjects including healthy non-obese volunteers and arginine-infused rats, were measured by N6E-EIA and 30K-RIA. The correlation between the two methods was evident ($r = 0.974$) but the absolute values of the IRG levels were slightly different (Fig. 4), that is to say, the values obtained by EIA were smaller than those obtained by RIA. The difference in IRG levels was also found in the comparison between N6E- and 30K-EIAs (Fig. 5). A number of papers have pointed out that the IRG in plasma of control subjects is heterogeneous and is present in at least two fractions, with a high molecular weight component (molecular weight >40000, big plasma glucagon) and material which is similar to pancreatic glucagon (molecular weight = 3485) (4). In these experiments, using the BPG and PG fractions obtained from nine plasma specimens, we obtained results showing that the N6E-immunoreactivity in EIA using peptide IV-enzyme conjugate was less against the BPG than the 30K-immunoreactivity in EIA using PG-enzyme conjugate, while both antisera showed a similar immunoreactivity against the PG fractions (Table IV). The difference between the plasma IRG levels obtained in the N6E and 30K systems may be explained by the difference of immunoreactivity against the BPG in plasma specimens to be tested. A low correlation (r = 0.661) obtained in the comparison between N6E- and 30K-EIAs was probably due to the use of only basal samples, in which BPG represents a high percentage of the total glucagon immunoreactivity. In fact, the correlation coefficient between N6E-EIA and 30K-RIA decreased from 0.974 to 0.853 when out of 25 specimens (Fig. 4), only the basal samples (n = 14) were compared. Neither antisera 30K nor N6E cross-reacts with gut GLI because they have a population of antibodies against the C-terminal portion of a glucagon molecule but none against the N-terminal and middle portions. However, the results obtained here show that the two antisera have a different specificity against BPG. As it is thought that the specificity of glucagon antiserum 30K is not only a function of C-terminal antigenicity alone but dependent upon the presence of the whole molecule (38), there might be a difference in the steric configuration between C-terminal peptide residues in the dissociated state and those in the native peptide hormone, and, although little is known about its structure, BPG might contain a conformation similar to the C-terminal portion of a glucagon molecule.

In the clinical samples, the N6E-EIA system was compared with the RIA using PG-specific antiserum G21, the main immunological determinant of which is the C-terminal peptide of glucagon (21-29) (13). After a glucose load, plasma glucagon, measured by EIA and RIA, fell in 12 normal subjects. The IRG levels measured by both assays during glucose tolerance test showed good agreement in a group of 10 patients. After arginine infusion, plasma glucagon increased in 6 normal subjects and 3 patients and the IRG values measured by N6E-EIA correlated well with those by G21-RIA. There were several poor correlations at low IRG levels in these tests between the two methods, and the difference could be due to the use of different standards since both assays were performed separately in two laboratories, using different standards of pork crystalline glucagon (15). But the possibility that the immunoreactivity of G21 against the BPG in plasma specimens is different from that of N6E cannot be excluded although it was not investigated in this study.

The present work demonstrates that plasma IRG in human subjects can be measured by the EIA using antiserum N6E with the C-terminal peptide (21-29)-enzyme conjugate and a good correlation exists between plasma IRG measured by N6E-EIA and the conventional RIA. The EIA method for glucagon can be regarded as potentially competitive with the popular established RIA technique in counts of sensitivity, specificity and reproducibility, although the glucagon assay itself has some problems to resolve concerning glucagon standards, immunoreactivity of antisera against BPG and the stability of glucagon molecules in plasma specimens.

References

1. Sundby, F.: Species variations in the primary structure of glucagon. Metab. Clin. Exp. 25, 1319-1321 (1976).

2. Valverde, I., Rigopoulo, D., Marco, J., Faloona, G.R., Unger, R.H.: Characterization of glucagon-like immunoreactivity (GLI). Diabetes 19, 614-623 (1970).

3. Srikant, C.B., McCorkle, K., Unger, R.H.: Characterization of tissue IRGs in the dog. Metab. Clin. Exp. 25, 1403-1404 (1976).

4. Valverde, I., Villanueva, M.L., Lozano, I., Marco, J.: Presence of glucagon immunoreactivity in the globulin fraction of human plasma ("Big Plasma Glucagon"). J. clin. Endocrinol. Metab. 39, 1090-1098 (1974).

5. Unger, R.H.: Glucagon and the insulin:glucagon ratio in diabetes and other catabolic illnesses. Diabetes 20, 834-838 (1971).

6. Lawrence, A.M.: Radioimmunoassayable glucagon levels in man. Effects of starvation, hypoglycemia and glucose administration. Proc. natn. Acad. Sci. USA 55, 316-320 (1966).

7. Hazzard, W.R., Crockford, P.M., Buchanan, K.D., Vance, J.E., Chen, R., Williams, R.H.: A double antibody immunoassay for glucagon. Diabetes 17, 179-186 (1968).

8. Young, J.D., Kraegan, E.W.: Simultaneous assay of insulin and glucagon in serum. Aust. J. exp. Biol. med. Sci. 46, 697-705 (1968).

9. Nonaka, K., Foa, P.P.: A simplified glucagon immunoassay and its use in a study of incubated pancreatic islets. Proc. Soc. exp. Biol. Med. 130, 330-336 (1969).

10. Heding, L.G.: Radioimmunological determination of pancreatic and gut glucagon in plasma. Diabetologia 7, 10-19(1971).

11. Unger, R.H., Eisentraut, A.M., McCall, M.S., Madison, L.L.: Glucagon antibodies and an immunoassay for glucagon. J. clin. Invest. 40, 1280-1289 (1961).

12. Iwasa, S., Ueno, H., Miya, T., Wakimasu, M., Kondo, K., Ohneda, A.: Enzyme immunoassay of pancreatic glucagon at the picogram level using β-D-galactosidase as a label. J. Biochem. 86, 943-949 (1979).

13. Ohneda, A., Watanabe, K., Wakimasu, M., Fujino, M.: Production of a specific antiserum by synthetic C-terminal fragment of glucagon. Horm. Metab. Res. 11, 463-467 (1979).

14. Iwasa, S., Kondo, K., Ueno, H., Wakimasu, M., Ohneda, A.: Enzyme immunoassay for determination of pancreatic glucagon in plasma. Horm. Metab. Res. 13, 14-17 (1981).

15. Ohneda, A., Kobayashi, T., Nihei, J., Iwasa, S., Kondo, K.: Measurement of glucagon in human plasma by enzyme immunoassay. J. Immunoassay 4, 339-349 (1983).

16. Fujino, M., Wakimasu, M., Shinagawa, S., Kitada, C., Yajima, M.: Synthesis of the nonacosapeptide corresponding to mammalian glucagons. Chem. Pharm. Bull. 26, 539-548 (1978).

17. Kenny, A.J.: Extractable glucagon of the human pancreas. J. clin. Endocrinol. Metab. 15, 1089-1105 (1955).

18. Garaud, J.C., Moody, A.J., Eloy, R., Gremier, J.F.: Unusual specificities of antibodies to glucagon-glutaraldehyde-albumin conjugates. Horm. Metab. Res. 8, 241-243 (1976).

19. Faloona, G.R., Unger, R.H.: Glucagon. In: "Methods of Hormone Radioimmunoassay", Eds. Jaffe, B.M., Behrman, H.R., Academic Press, New York, pp. 317-330 (1974).

20. Ohneda, A., Ishii, S., Horigome, K., Yamagata, S.: Glucagon response to arginine after treatment of diabetes mellitus. Diabetes 24, 811-819 (1975).

21. Sakurai, H., Kuzuya, H., Kurahachi, H., Fukase, M., Imura, H.: Glucagon radioimmunoassay. Jap. J. Nucl. Med. 10, 135-136 (1973).

22. Kitagawa, T., Aikawa, T.: Enzyme coupled immunoassay of insulin using a novel coupling reagent. J. Biochem. 79, 233-236 (1976).

23. Iwasa, S., Kondo, K., Miya, T., Takeda, K.: Enzyme immunoassay of a β-adrenergic agent using β-galactosidase as label. Immunopharmacol. 1, 3-12 (1978).

24. Woolen, J.W., Walker, P.G.: The fluorimetric estimation of N-acetyl-β-glucosaminidase and β-galactosidase in blood and plasma. Clin. chim. Acta 12, 647-658 (1965).

25. Weir, G.C., Turner, R.C., Martin, D.B.: Glucagon radioimmunoassay using antiserum 30K: Interference by plasma. Horm. Metab. Res. 5, 241-244 (1973).

26. Imagawa, K., Nishino, T., Shin, S., Uehata, S., Hashimura, E., Yanaihara, C., Yanaihara, N.: Production of anti-glucagon sera with a C-terminal fragment of pancreatic glucagon. Endocrinol. Japon 26, 123-131 (1979).

27. Assan, R., Slusher, N.: Structure/function and structure/immunoreactivity relationships of the glucagon molecule and related synthetic peptides. Diabetes 21, 843-855 (1972).

28. Jacobsen, H., Demandt, A., Moody, A.J., Sundby, F.: Sequence analysis of porcine gut GLI-1. Biochim. biophys. Acta 493, 452-459 (1977).

29. Thim, L., Moody, A.J.: The primary structure of porcine glicentin (proglucagon). Regulatory Peptides 2, 139-150 (1981).

30. Van Weemen, B.K., Schuurs, A.H.W.M.: Immunoassay using hapten-enzyme conjugates. FEBS Lett. 24, 77-81 (1972).

31. Van Weemen, B.K., Schuurs, A.H.W.M.: The influence of heterologous combinations of antiserum and enzyme-labelled estrogen on the characteristics of estrogen-enzyme-immunoassays. Immunochemistry 12, 667-670 (1975).

32. Ohneda, A., Sato, M., Yamagata, S.: Insulin response to tolbutamide-glucose load in normal and diabetic subjects. Tohoku J. Exp. Med. 114, 171-182 (1974).

33. Ohneda, A., Sato, M., Matsuda, K., Yanbe, A., Maruhama, Y., Yamagata, S.: Plasma glucagon response to blood glucose fall. Tohoku J. Exp. Med. 107, 241-251 (1972).

34. Morgan, C.R., Lazarow, A.: Immunoassay of insulin: Two antibody system. Diabetes 12, 115-126 (1963).

35. Kato, K., Hamaguchi, Y., Fukui, H., Ishikawa, E.: Enzyme-linked immunoassay. I. Novel method for synthesis of the insulin-β-D-galactosidase conjugate and its applicability for insulin assay. J. Biochem. 78, 235-237 (1975).

36. Kikutani, M., Ishiguro, M., Kitagawa, T., Imamura, S., Miura, S.: Enzyme immunoassay of human chorionic gonadotropin employing β-galactosidase as label. J. clin. Endocrinol. Metab. 47, 980-989 (1978).

37. Ishikawa, E.: Enzyme-labelling of antibodies. J. Immunoassay 4, 209-327 (1983).

38. Vinik, A.I., Hardcastle, A.: Structure antigenicity relationships of glucagon and related peptides. Horm. Metab. Res. 6, 396-399 (1974).

RECENT ADVANCES IN ISOELECTRIC FOCUSING THEORY, TECHNIQUE, AND APPLICATIONS OF VALUE IN IMMUNOLOGY AND RELATED DISCIPLINES

J. H. Jackson, Jr.
Department of Psychiatry M-003, School of Medicine, University of California at San Diego, La Jolla, California 92093, U.S.A.

I. Introduction

A) General approach of the review

This review focuses upon artifacts likely to be encountered during isoelectric focusing (IEF) on polyacrylamide gels. Its single goal is to present recent developments in, and improvements of, polyacrylamide gel isoelectric focusing (PAGIF), and it has been divided into three parts. Although this centers around issues of immunological importance, it is a general survey suitable for anyone who uses the technique. In addition to the PAGIF method, it contains important new considerations about column electrofocusing of value to the protein chemist, and has been designed to present new theories and techniques for the practical solution of commonly encountered problems and to suggest ways in which they can be applied in clinical situations. Part I presents recently developed theories about the isoelectric point (pI) that extend earlier IEF concepts. In general, the revised extensions recognize the dynamic nature of macromolecules and thus view the pI as a variable parameter which reflects changes in the shape of a molecule and its interaction with other molecules. Part I also identifies PAGIF as the most commonly encountered form of IEF, and thus the central IEF method around which the review will pivot. Part II takes an in-depth look at the most common problems encountered during PAGIF and presents recent techniques which

Immunoassay Technology, Volume 1

may help to reduce or eliminate these problems. The emphasis, which is only marginally centered around immunology in part I, is more clearly centered around improvements of special interest to the immunologist in part II. However, the techniques described are of equal importance to all who use PAGIF and who will encounter the same difficulties, but to varying degrees.

In part III the emphasis is almost entirely centered around immunology. Modifications and improvements described here are placed within the framework of clinically encountered immunological situations. This part centers around those ideas presented in parts I and II which can be included into techniques of proven value in the clinical laboratory. Thus the progression will have moved from a general, non-clinical, non-immunological tone to a less general, more immunological but non-clinical tone to a focus which is almost entirely geared to clinical immunology.

This review focuses entirely upon PAGIF artifacts and techniques to eliminate them and is not intended as a general, revised survey of the IEF technique, which information is available elsewhere (1-5).

B) Historical background and theory

Isoelectric focusing has become a widely used analytical tool since it was first introduced about two decades ago, and its value is rapidly increasing due to new theoretical concepts and improved technical advances. Although the column technique was once the most widely used and encountered form of IEF, today polyacrylamide gel isoelectric focusing is most frequently encountered, particularly as an analytical tool in immunology, genetics, and in many other disciplines.

Isoelectric focusing is based on the observation that a protein placed in a pH gradient within an electric field will migrate until it reaches a pH at which there is no net charge on the molecule (1). Upon reaching this pH, called the isoelectric point or pI, the protein will cease its migration and become concentrated or "focused". Thus, isoelectric focusing is valuable for purifying and concentrating proteins. Recent theoretical extensions of IEF acknowledge that the motion of molecules, particularly conformational changes (6), ligand and drug interactions (6-8), interactions with allosteric effectors (7), and macromolecular interactions (6-8) markedly alter the behaviour of proteins studied by this technique.

C) Revised theories, new techniques and practical applications

Many uses of IEF encountered in the literature involve modifications of the original IEF theoretical ideas, as expressed most succinctly by several workers (6-10), who also seem to have taken into account the widespread potential applications of IEF in biochemistry, pharmacology and other disciplines (6, 7). They note that systems experiencing rapidly reversible, carrier ampholyte-induced macromolecular isomerizations, association-dissociation reactions, or pH-dependent conformational changes can display well resolved bimodal or multimodal focusing patterns in which the peaks correspond to different equilibria compositions rather than to separated macromolecular species (7). Although they attribute most multimodal peaks to macromolecular interactions with the ampholyte buffers, they also note that the theory applies to the detection of specific interactions between enzymes and other proteins, with each other, with ligands such as allosteric effectors, and with drugs, and present two equations in which the complex formed by interacting molecules displays a pI value which is intermediate between that of either of the interacting species (7). The essential difference between the two equations is that one results from the interaction between two amphoteric molecules whereas the other results from the interaction between one amphoteric and one non-amphoteric molecule. Each interaction is theoretically thought to yield its own distinct IEF spectrum (7).

Three possible explanations have been proposed (6) for such patterns: 1) the ampholyte buffers can bind directly to the macromolecule of interest, 2) proteins may experience several sequential pH-dependent conformational transitions as the bases are protonated and, 3) both the above mechanisms may occur since neither is mutually exclusive of the other. We have recently presented evidence for extensions of these mechanisms applicable to comparing the spectral patterns of purified and crude systems (11). Our studies indicate that macromolecular interactions with each other should be included in the above mechanisms as proposed elsewhere (7). According to this extension, the removal of certain interacting molecules from the environment modifies the pI and spectral profile of the molecule of interest. Indirect interactions are also possible in which a molecule that normally binds the protein of interest interacts with a third molecule which does not typically bind that protein. This may result in a multimolecular complex involving three or more interacting molecular species which can include ligands or drugs. These ideas are of value for interpreting studies of immune complexes.

Altered spectral patterns need not result uniquely from interactions between molecules, but can also result from altered conformational states induced by solvent effects, electrostatic effects, or other factors known to induce conformational changes (12). Rather than causing interactions, these effects simply alter the conformation to such an extent that different amino acid side chains are brought to the surface of the molecule where they are exposed to the outer environment. The resultant conformational change may alter the degree of protein-protein interactions.

These theories are crucial for understanding the various IEF spectra that emerge during variations of the technique. Thus, the pI's one observes during an experiment may not represent that for the protein of choice, but rather an altered index that reflects macromolecular interactions, conformational changes, drug interactions, etc. This fact has increased the value of the IEF technique in many disciplines, including immunology, and has added to the information that can be extracted from IEF data.

D) Isoelectric focusing in polyacrylamide gels (PAGIF)

Although new theoretical IEF concepts (6, 7) have arisen over the past decade, most have only been tested by PAGIF procedures (13-21) due to the extensive use and relatively low cost of gel procedures, PAGIF gels being important as analytical tools. Were it not for gel electrofocusing procedures, the value of the IEF technique as an analytical tool would be greatly diminished. Protein bands of significance to the clinician or laboratory analyst are often so close together that adjacent bands would unavoidably mix during elution from a column, rendering the column procedure much less useful. PAGIF procedures have several advantages over column electrofocusing as analytical tools. One major advantage is cost. Except for the recently reported minicolumn technique (22), columns are more expensive than gels and also require more time to focus, usually between 12 and 48 hours depending on the protein, as compared with 2-5 hours for PAGIF, although longer periods of focusing can be encountered (23). As mentioned above, the most significant factor against the use of columns as analytical tools is the remixing of adjacent bands during elution. This is an important point. Even improper staining techniques can be a source of error when analyzing PAGIF bands causing one to miss important proteins unique to the spectrum (24) but this can be eliminated. Thus, the gel electrofocusing techniques are usually the only practical IEF techniques for analytical purposes.

A literature survey on isoelectric focusing quickly reveals the widespread value and application of the IEF technique. Except for protein purification, the PAGIF method is the electrofocusing method of choice and has found widespread application in genetics (13, 25, 26), bacteriology (27), the direct analysis of tissue for disease detection (28, 29), the analysis of proteins in biological samples (23, 30-33), desorbing substances tightly bound to affinity gels (34), cell fractionation (35-37), forensic science (38-40), and a host of other uses.

Despite their value as analytical tools, gel techniques also have a number of drawbacks (41-49). Of most concern are the numerous artifacts associated with essentially every gel technique, including PAGIF. Those associated with "cathodic drift" are of most concern and will be considered under "Dealing with cathodic drift and other artifacts". A second problem unique to gels is that very large proteins such as IgM, IgA, and some antigen-antibody complexes are too large to enter the PAGIF gel (17, 50, 51). This factor will also be considered later. Even with these shortcomings, PAGIF is superior to column IEF and other methods of analyzing body fluids.

II. Methods

A) Improved fixation methods and related techniques

Staining techniques. The choice of staining method is extremely important to the interpretation of PAGIF data. Staining is the most direct method to detect focused proteins, but staining IEF gels is not as straightforward as staining electrophoretic gels. The ampholyte buffers used to establish the pH gradient bind to the two most commonly used dye stains, Coomassie blue G-250 and Coomassie blue R-250, resulting in a significant amount of background staining. Several measures can be taken to prevent artifacts which result from background staining, the most common steps being to remove ampholytes and fix the protein in the gel before staining with Coomassie blue R-250 (52, 53) or to stain with Coomassie blue G-250 in perchloric acid (54, 55). Both procedures can produce inconsistent results. To overcome artifacts caused by staining, one must either modify the staining procedure, use a different type of staining dye, or use another approach to visualize proteins. Each alternative has been attempted with varying degrees of success.

A modified staining technique has recently been developed which avoids the use of chemical fixatives while serving as a one-step method to elute and renature proteins (24). In this study Coomassie blue G was found to be superior to either Coomassie blue R or fluorescamine. Background staining was minimized by using 0.03% Coomassie blue G dye. Glycerol (5%) enhances and Triton X-100 prevents staining. Both pH and the molecular weight of the proteins were identified as major factors in Coomassie blue G-250 stains. Maximum staining is obtained at pH 9.0.

Artifacts caused by background staining of ampholytes can be reduced or eliminated by using fast green which binds proteins but not the ampholyte buffers (56, 57). Maximum staining can be achieved within 5 minutes and bands become visible within 3-6 hr of destaining (58). However, fast green is not as sensitive as Coomassie blue G-250, staining 0.5-8.0 micrograms actin as compared to 0.1-2.5 micrograms actin for the Coomassie dye. The linear region of dye binding is larger for fast green than for Coomassie blue G-250.

An alternative to modifying a standard dye approach or to using less common dyes is to use the immunoperoxidase stain (58, 59). The immunoperoxidase method is an immunoenzymatic procedure used to distinctly stain and identify the species of ameba present in the cerebrospinal fluid (CSF) in cases of soil ameba infection. Although usually applied to formalin-fixed paraffin sections (60), it can be used to stain nitrocellulose protein blots of PAGIF gels (56). The nitrocellulose blot is soaked for 10 minutes in a solution containing 10 mg 3'-amino-9'-ethylcarbazole dissolved in 6 ml dimethylsulphoxide (BDH).

<u>Fixation methods</u>. Once focused in the PAGIF gel, proteins must be fixed to maintain sharp bands and to remove the ampholyte buffers (57). Fixation often results in artifacts. The most commonly encountered fixation methods have already been reported (61, 62). The former method (61), precipitates immunoglobulins with 18% sodium sulfate which has been added to the labelled antigen and the washing solution. It is useful to assess antibodies to haptens since larger antigens are often

precipitated. The second method precipitates immunoglobulins with 18% sodium sulfate then fixes the gel with glutaraldehyde, thus eliminating problems caused by the precipitation of larger antigens (63). However, this method can damage the antibody binding ability. A second compound, dimethylsuberimidate, can replace glutaraldehyde as a chemical fixative. Both fixatives destroy the ability of antibodies to bind antigens, even at fixative concentrations too low to accommodate the complete fixation of protein bands (63).

Two methods have recently been introduced to overcome fixation artifacts, by eliminating the chemical fixative (51) used in one method (62) and modifying a second antibody precipitation technique (64) in an effort to eliminate those artifacts caused by chemical fixatives. The modified technique soaks the gel in 18% sodium sulfate for 10 minutes, aspirates the gel and incubates it overnight in antiserum at room temperature followed by three washes in phosphate buffered saline (PBS) over the next 24 hours. After this process, thyroglobulin (Tg) antibody bands are sharper and nonspecific binding to two normal human sera is essentially eliminated. Sodium sulfate is thought to enhance the efficiency of antiimmunoglobulin precipitation by decreasing the solubility of immune aggregates (51, 65).

A second procedure used to overcome the effects of chemical fixatives preincubates the precipitated antibodies with radiolabelled antigen before fixation (63). Prefixation treatment removes all low molecular weight, amine containing compounds such as ampholytes, electrolytes, and urea from the gel under conditions that prevent antibodies from diffusing out. This increases the sensitivity of protein detection by eliminating the interference of ampholytes. The prefixation treatment consists of agitating the gel in 21.5% sodium sulfate for 30 min at 30°C and washing three times successively for 30 min each at room temperature. The gel can now be overlaid with radiolabelled antigen in an 18% sodium sulfate solution then washed and fixed. Gels pretreated as described result in spectrotypes

(ST) which are dependent only upon the retention of antibody in the gel, not on the fixative concentration. As glutaraldehyde is reduced from fixing to poorly fixing concentrations (0.0033%), the sensitivity of autoradiographic detection approaches (but never reaches) that of the control. Heterogeneity and band intensity change as a function of fixative concentration (63). Glutaraldehyde fixation differentially inactivates neutral bands whereas alkaline bands are more resistant. It is concluded (63) that electrofocused spectrotypes visualized after glutaraldehyde fixation reveal those bands most resistant to inactivation and is not truly representative of antibody heterogeneity.

B) Techniques combined with conventional and ultrathin PAGIF

Ultrathin PAGIF. One of the most promising PAGIF techniques is the ultrathin gel method. This procedure has been applied in a number of studies of body fluids (66-68). Typical PAGIF gel thickness is between 1.5-3mm (2, 44, 69, 71). A number of recent studies have employed 1 mm gels (56, 70) but even thinner gels (ultrathin gels) may be ideal for PAGIF studies. In the past, ultrathin gels have been avoided because they are difficult to handle (70), but this problem has been overcome by polymerizing acrylamide on cellophane (70, 72). The polymerized gel adheres firmly to the cellophane during all subsequent focusing, staining, and destaining procedures and subsequently can be dried to the cellophane so as to provide a permanent record of the results for future reference.

Several ultrathin gel thicknesses have been employed, including 0.2 mm (67, 68), 0.55 mm (56), 0.12 mm, 0.24 mm and 0.36 mm thickness (70). In each case, the ultrathin procedure has demonstrated significant advantages over conventional gels and has eliminated many problems associated with PAGIF artifacts. In particular, ultrathin gels allow for a more efficient heat transfer, the application of high field strength, improved resolution, shorter focusing time and faster staining and destaining (70, 73, 74). These improvements should decrease or eliminate the likelihood of cathodic drift (75).

Recently, two modifications of the ultrathin technique have been presented (70, 76) where one (70) compared the electrofocusing patterns obtained using three different ultrathin gel thicknesses, including 0.12 mm, 0.24 mm and 0.36 mm. The ultrathin gels universally and reproducibly increased the sharpness of protein bands but the patterns obtained with 0.12 mm gels were distorted, particularly in the alkaline pH range. Band distortions were attributed to electroendosmosis probably caused by unfavourable ratios of gel volume to surface area and the interaction of proteins with the cellophane support. These distortions could not be eliminated by increasing the ampholyte concentration to 4%. Thus, there appears to be an optimal range of PAGIF gel thickness between 0.12 mm and 1.0 mm. An added value obtained with the ultrathin procedure is that adjacent proteins which usually focus as a single band can be resolved into separate bands (70). Staining and destaining procedures were considerably shorter for the ultrathin procedure but could be further shortened by applying quick staining procedures that minimize interaction between stains and ampholytes (54). Even with the shortened destaining time associated with ultrathin gels, further time reductions may be desired. The longer destaining is continued, the greater the chance that protein zones will gradually lose their stain (77). No background staining is observed with ultrathin gels. Thus, pretreatment with trichloroacetic acid (TCA) to remove ampholytes (1, 44) can be eliminated although it does significantly improve the sharpness of the protein bands.

<u>Two-dimensional gels</u>. Isoelectric focusing can be combined with sodium dodecyl sulfate (SDS) gel electrophoresis into a two-dimensional (2-D) procedure which separates proteins by exploiting two of their fundamental physical parameters, the isoelectric point and the molecular weight. This procedure was further developed simultaneously and independently (78, 79),

and has been widely used over the past 5 to 10 years. Recently, it was sought to eliminate (24) the use of chemical fixatives and to obtain an optimal staining method for visualizing gel spots during the PAGIF procedure, as mentioned above. The ultrathin technique was recently applied (76) to the two-dimensional procedure as a second modification of the ultrathin PAGIF technique. Using SDS electrophoresis as the second dimension, the two-dimensional procedure with ultrathin gels was found to be sufficiently sensitive to detect small amounts of proteins, including impurities (less than 0.1 microgram protein/band) in M.W. markers. Molecular weight determinations by this method were accurate, protein bands were extremely sharp and background staining was minimal (76). The two most critical steps in the 2-D technique with ultrathin gels are the formation of reproducible exponential pore gradients and the method of applying the sample to the gel. For high reproducibility, gradients must be allowed to form slowly. The best results are obtained when liquid samples are applied in a pocket pre-cast in the gel and gel strip samples are applied in a trench (76). This procedure allows for the detection of alkaline bands which would normally be lost during electrofocusing on conventional PAGIF gels, including other 2-D procedures.

While alluding to the subject of the loss of alkaline bands, reference should be made to another recent procedure which reduces the loss of alkaline bands. Although the PAGIF procedure has been described for narrow pH gradients in the alkaline range, problems are often encountered in attaining the degree of resolution typical within other pH ranges (140). Difficulties are thought to arise because 1) the procedure is unsuitable for resolving protein mixtures which contain a significant amount of protein with pI's outside the buffering range of the system, 2) the low field strength which can be attained in the far alkaline range, and 3) the significant sieving effects in the polyacrylamide (91). A two-stage PAGIF procedure has been developed (91) to overcome some of these problems, where the sample is electrofocused between pH 7 and pH 9 for 4 hours and the gel containing the focused proteins is cut so as to remove the anodal portion of the gel (at about pH 7.0) which is subsequently discarded. The second step is performed between pH 8.5 and pH 10.5. This procedure allows for good resolution of cationic gamma globulins in serum and the application of a high field strength. Improved results are attributed to the elimination, during the second stage, of carrier ampholytes in the acidic pH range which are of low and uneven conductivity (44) and thus are irrelevant to the procedure.

A major concern during this two-stage procedure is preparing a suitable polyacrylamide with an average pH above 8. Since the pI of gamma globulins appears to increase with increasing molecular size (80, 81), low acrylamide gel concentrations are required for maximum separation of the most cationic serum proteins.

Enzyme linked immunosorbent assay (ELISA). The most commonly encountered means to detect protein or immunological reactivity are stains (56, 57), radioimmunoassay (51, 107), immunofixation (105), protein blotting (86) and enzyme linked immunosorbent assay (ELISA) (82, 83). The latter method, ELISA, which has been described (82), is particularly useful for studying immune complexes dissociated during isoelectric focusing since it can be used to recombine isolated antibodies to a solid phase so as to demonstrate immunological reactivity (84). The technique involves three steps. First, free antigens are physically attached to a solid phase. Isolated antibody is then allowed to interact with the bound antibody, resulting in an antigen-antibody complex bound to a solid phase. Subsequently, it is necessary to detect the bound antibody. This technique has been successfully used (84) in combination with chromatofocusing to demonstrate immunological reactivity. Other studies have combined it with other IEF procedures to detect immunological activity (56, 85).

Protein blotting. Upon completion of the PAGIF procedure, protein bands can be transferred from the gel to immobilizing matrices of modified cellulose, a process known as protein blotting. The blotting procedure can also be used to detect RNA (RNA blotting) or DNA (DNA blotting) and is thus valuable for detecting protein-protein or protein-ligand interactions and for immobilizing polypeptides so that they can be subjected to immunological or biochemical analyses (86, 167). Briefly, the process involves placing prewetted nitrocellulose on top of the focused gel taking care not to move it once it touches the gel surface. A second piece of prewetted nitrocellulose can be placed on the bottom of the gel. The entire gel-nitrocellulose assembly is placed between two pieces of prewetted Whatman 3MM paper and two glass plates and allowed to remain for two hours. At the end of this period the sandwiched gel unit is disassembled and the top and bottom units are rinsed with water or buffer prior to use. The PAGIF gels are recovered intact and can be stained to detect protein (56, 87). An electrophoretic blotting process (88) is thought to speed the transfer process (87) and retain the sharpness of focused bands (89), but the rapid transfer of macromolecules from IEF gels (about 60 minutes) with sharp bands, by a non-electrophoretic method, has been reported (56), showing that the presence of SDS in gels reduces the transfer time required and identifying the following factors as contributing to the speed of protein transfer: a) gel thickness, b) degree of cross-linking, c) manner of sample application, and d) the tendency for proteins to focus on the surface of IEF gels.

Replicate blots can be made since both the top and bottom of the PAGIF gel can be blotted (90). Subsequently, each blot can be used for different purposes. For example, one can be stained while the second can be used for ELISA, binding studies etc. Furthermore, as mentioned above, the original gel remains intact once blotting is completed, and thus can also be used for whatever purpose is required. When preparing replicate blots, one should keep in mind that the top and bottom of the

PAGIF gel do not stain comparably, but the ratio of protein on the top and bottom blots is constant for a given acrylamide gel thickness (56). Some blotting techniques may result in a loss of band resolution during transfer from agarose gels to nitrocellulose (59).

Protein blotting techniques are valuable tools in that they require small amounts of reagents, relatively little time, can be overlaid with a similar variety of probes as can the original polyacrylamide gel, have a sensitivity up to 1 pg of protein, and can be reused by removing the blotted protein with low pH or with protein denaturing conditions such as SDS, urea, or potassium thiocyanate.

C) Dealing with cathodic drift and related artifacts

Probably no topic discussed in this review is more important to immunologists than the issue of cathodic drift. This phenomenon is a distortion or decay of the pH gradient characterized by acidification at the anode, accumulation of ampholytes at the cathode, and formation of a plateau in the neutral pH region (75). Overcoming problems encountered when electrofocusing immunoglobulins often adds to the decay of the pH gradient. In particular, the low percentage acrylamide gel matrix, usually around 5%, and the high voltage applied for several hours, tend to aggravate cathodic drift and the convection flow of heat, ions and water (63). Several procedures have been devised to overcome the various aspects of this complex problem. One uses a two-stage PAGIF method (91), a second covalently attaches polyacrylamide layers to glass so as to give mechanically stable gels and generate stable, reproducible pH gradients (63). A similar technique uses immobilized ampholytes called Immobilines (92, 93).

Stable pH gradients can be achieved by replacing carrier ampholytes with amino acid electrolyte solutions whose pI's match the pH values at the ends of the gradient (63, 94, 95). High voltages can be used to overcome the low conductivity of the amino acids (94, 95). In such a situation, 2% ethanolamine or 0.05 M lysine must replace NaOH as the alkaline solution because amino acid hydrochlorides are not suitable for isoelectric focusing in the presence of NaOH (63).

One group of workers (94, 95) have replaced carrier ampholytes with electrolytic solutions of the free base of amino acids, which overcomes the instability of the pH gradient as long as certain basic considerations are kept in mind. In another study (63), either 0.0043% or 0.05M glycylglycine served as the anode solution and 0.05M lysine or 0.1% ethanolamine as the cathode solution. However, one should recall that amino acid hydrochlorides which have been adjusted to their pI values with base are not suitable for isoelectric focusing. IEF theory dictates that amphoteric compounds with good conductance at their pI's or small differences between the pI and pK of the ampholyte most proximal to its pI (i.e. $pI-pK_p$) serve as suitable carrier ampholytes (96). This includes some amino acids such as histidine and glutamate but excludes others such as glycine. Increasing the ionic strength is theoretically believed to decrease the likelihood that a natural pH gradient will form (1). However, it has been demonstrated (97) that 0.1 M KCl actually increases the formation of a pH gradient when poor electrolytic materials such as glycine are used to form the gradient. Furthermore, the increased ionic strength permits both cationic and anionic molecules to migrate to their pI's at higher rates than does low ionic strength, allowing for shorter focusing times. Even at high ionic strength, the molecular migration to their pI values proceeds asymmetrically, being higher for anionic than cationic components (97). This is an important consideration because some molecules such as IgA or IgM must be loaded at the anode in order to penetrate the PAGIF or agarose gels (98).

Cathodic drift is of major concern to immunologists, particularly those studying multiple sclerosis (MS), because alkaline IgG's tend to be characteristic of some immunological disorders (50, 99-106). Thus, an accurate assessment of proteins within the alkaline region between pH 8.5 and pH 9.5 is often essential. Reference was made earlier to the loss of alkaline bands and to a recent two-stage procedure (91) which reduces the loss of those bands (see "two-dimensional gels"). A useful extension of this technique involving minicolumns (22) is proposed under "Purification methods" in this review. The two-stage procedure eliminates artifacts attributed to the acidic carrier ampholytes which are of such low and uneven conductivity as to generate excess heat during the electrofocusing procedure. Two other factors contributing to artifactual bands within the alkaline pH range are CO_2 (23, 41) and the non-equivalence of the cathodic and anodic ends of the gel (98).

When performing the PAGIF procedure, one should be aware that cathodic and anodic ends of the gel are not equivalent. In a report on composite gels, it was noted (98) that IgA and IgM samples applied to the cathodic end of the gel did not enter any gel system tested, including polyacrylamide, agarose, or a composite of the two. This was thought to be due to the properties of the immunoglobulins since IgG migrated to its pI irrespective of which end of the gel it was loaded. It is more likely due to the combined properties of the immunoglobulins and the gel systems. In one study (94) cathodic drift is attributed to discontinuities in the pH gradient between the electrode terminals, whilst in another (97), discontinuities are viewed as the barriers to protein migration. Any number of factors might contribute to the formation of these barriers, two of which are well documented. One case is that of 2-mercaptoethanol (2-ME) (108) which ionizes at the basic end of the PAGIF gel and is electrophoretically driven along the entire pH gradient. As it migrates, it sweeps focused carrier ampholytes away along its path. Thus, 2-ME strongly interferes with pH gradients and protein mobilities, particularly in the alkaline region where carrier ampholytes are driven away from their pI position and lost in the cathodic chamber. Although proteins may add to the gradient distortion due to their buffering capacity (109), it is noted (108) that their contribution to the distortion of the pH gradient is negligible compared to that resulting when 2-ME is present.

A second well documented contribution to distortion of the pH gradient is CO_2 (23, 41). Studies of serum and CSF reveal that alkaline immunoglobulins are synthesized by the CNS immune system of MS patients (50, 100-104). The pI of these proteins falls between 5.9 and 9.3, but some fall within the range of 8.3-9.3 (50, 100, 101, 103). It has been shown (41) that CO_2 in the air and in the gel interferes with the separation and pI determination of alkaline proteins, particularly those above pI 8.2-8.3. Citing references which present the pI's for ribonuclease, cytochrome c, and a variety of other enzymes, it is noted that CO_2 can lower the pI value by as much as 1-2 pH units and add to the cathodic drift of the system. The CO_2 effect appears to occur in gels only, not on density gradients in columns, and, subsequently, similar CO_2 induced pI shifts in bovine liver histones were observed (23). Increasing the time of focusing adds to cathodic drift and it was noted that the pI of bovine histones focused for 20 hours in the presence v. absence of CO_2 were markedly different. To eliminate CO_2 induced cathodic drift, experiments in a nitrogen atmosphere were performed.

Several other problems associated with cathodic drift are mentioned in various reports. Among the solutions to these problems are the two-stage method of electrofocusing (91), the use of free base amino acids to form the pH gradient (63, 94, 95), and the use of materials known as Immobilines (92, 93) or of stabilized gradients attached to glass (63). The use of the free base of amino acids can add to electroendosmosis (63), that is to say, the migration of electrolytes between electrodes induced by fixed charges within the gel matrix. Resolving problems associated with cathodic drift is a complicated subject which should be the focus of an updated review of its own.
In the meantime, simple precautions such as using dithiothreitol instead of 2-ME, stabilizing pH gradients by attaching ampholytes to glass, use of Immobilines, performing the PAGIF

in a nitrogen atmosphere, always loading protein samples at the anodic end of the gel, and using amino acids to form the pH gradient will be helpful.

D) Purification methods

Although PAGIF is an extremely valuable analytical tool for separating individual proteins, it has not proven as valuable in scaled-up preparative procedures. Whereas many analytical techniques have been described, the available preparative techniques are comparatively limited and involve electrofocusing in density gradients or on gels (110-112). Preparative electrofocusing techniques have been inhibited by the high cost of ampholytes and their tendency to complex with a number of macromolecules so as to modify the spectral pattern (9, 44, 113-115).

A number of new preparative approaches rely on isoelectric focusing in granular gel beds such as G-75 Sephadex as introduced and developed by Radola (42, 47). Although this approach is valuable, the granular surface of gel beds is often wavy and blind preparative slices made in a straight line across the gel slab will often cut across multiple bands so as to result in the remixing of separated proteins. Fluorescence of the Servalyt ampholytes (available from LKB) was used (116) with a U.V. light source to locate and mark bands which can subsequently be separated and the purified protein removed from the Sephadex by filtering fractions through a small column plugged with cotton and sand. Eluted protein can be concentrated by lyophilization. Separation achieved by electrofocusing in granular beds equals or exceeds that obtained with columns or polyacrylamide gels, but the author found this procedure cumbersome and impractical. Granular Sephadex G-75 beds have also been used (104) to isolate immune complexes on a preparative scale, but few details relating to the success as a preparative procedure are presented. Polyacrylamide gels are superior to all electrofocusing methods for separating proteins but recovery of focused protein is difficult and the amount of protein that can be isolated is limited.

Furthermore, ultrathin gel studies (70, 76) show that band resolution is reduced by increasing the gel thickness so as to accommodate a preparative procedure. A recent report describes a multipurpose technique that can be used for electrophoresis, isoelectric focusing, or a combined 2-D procedure (117), and which is reported to be useful for either preparative or analytical procedures.

Agarose is useful for preparative electrofocusing. Its major drawbacks are the limited amount of sample that can be applied and the tendency to encounter electroendosmosis (118). This phenomenon has been defined along with cathodic drift elsewhere in this review. This problem has been circumvented (118) by using Isogel (Marine Colloids, Rockland, ME), a specially developed 'electroendosmosis-free' agarose gel, from which it is concluded that agarose is a sensitive technique for high resolution on a preparative scale, useful to isolate antibody bands either in a single step or in conjunction with a prior step in a crude IgG preparative procedure. However, single agarose bands can be resolved into multiple bands by PAGIF (118). It may be possible to employ a scaled-up version of the composite gel technique (98) as a preparative procedure, but this approach has yet to be tested on a preparative scale. Resolution of a single polyacrylamide band into multiple agarose bands has also been reported (152), so single and multiple bands are clearly a function of the proteins involved and the gel system used to separate them.

Despite reports suggesting the value of various IEF gel methods in protein purification (104, 116-119), no preparative gel procedure is superior to a successful column technique in terms of the amount of protein one can isolate. Gels tend to give much better separation and significantly reduce the chances that adjacent bands will remix once separated (118). By contrast, even though it is all but impossible to prevent remixing of separated bands in columns, the amount of protein that can be purified on columns far exceeds the amount possible on gels. Whereas only milligram quantities of total protein can be applied to gels (2), gram quantities of total protein can be applied to columns (1), and milligram quantities of a single protein can be recovered (1, 22). Furthermore, a preparative procedure always involves the use of as much protein as possible, a factor which decreases the sharpness of protein separation in gels. In addition, numerous artifacts are associated with gel procedures that are not encountered in columns (41, 44). In fact, virtually every artifact discussed in this review is associated exclusively with various gel procedures.

Any purification protocol will combine different techniques in order to effect a successful purification, but even if all the steps necessary to produce a totally homogeneous protein are known, the order of combining each step is important. In electrofocusing one might exploit the same property twice by focusing over a wide pH range then refocusing in a narrow pH range (1, 91). However, from the author's experience, literature tends to overlook the possibility of electrofocusing preparative amounts of crude samples over a wide pH range on columns and then refocusing the same material in a narrow pH range on preparative gels. This approach will eliminate many proteins in the first step so as to allow sharper focusing when the preparative gels are used while allowing one to take advantage of the unique characteristics and advantages of each IEF procedure. In some respects this procedure is comparable to a reported two-stage method (91) since the acidic carrier ampholytes which cause uneven conductivity and contribute to Joule heating can be eliminated after the column step. Subsequently, alkaline proteins can be refocused in a narrow, alkaline pH gradient in a second stage on a gel. One must remember to use 10% ampholytes in the column step (1) so as to have sufficient carrier ampholytes to last through both stages. Because of the large amount of carrier ampholytes required for the columns it is more economical to use the 20 ml minicolumns (22) or 10 ml microcolumns discussed below.

It is generally assumed that IEF columns give the same focusing pattern as IEF gels (21). This assumption is not always accurate (22, 23, 41, 97). Studies (120) with the monooxygenase tyrosine hydroxylase (T-OH) suggest that a new minicolumn technique gives a different focusing pattern than the PAGIF gel technique (121). Typical commercially available 110 ml and 440 ml columns cannot be used in some monooxygenase studies at all (122). Only one report of success exists (123), and details are sketchy. Neither can one assume that the pI of the crude material will be within ± 1.0 pH unit of the purified material (11). We have shown that the pI of purified tryptophan hydroxylase (TP-OH) from the rabbit hindbrain is almost 7 pH units higher than that displayed by the crude homogenate (11).

A number of important considerations need to be noted when comparing column and gel procedures. In addition to the non-equivalence of column and gel spectra, one major drawback with the column technique has gone unnoticed until recently. Electrofocusing on columns can result in the unfavourable loss of a large amount of protein activity. When encountered, such losses are only seldom reported (22, 122, 124, 125). The cause of these losses is unclear but they appear to result from phenomena associated with column size (22). A 5%-30% loss of adenylate kinase (AK) activity was noted (124) when electrofocusing the enzyme from several different sources on 110 ml columns but no significant activity loss was reported when different isozymes were electrofocused on 20 ml minicolumns (166). Major losses of AK activity were noticed when using 110 ml and 440 ml columns (126) but these losses disappeared completely when identical AK isozymes (i.e. from the same source) were electrofocused on 20 ml minicolumns, an observation which has recently been confirmed (22). Whatever the cause, the experimental outcome is not unlike the increase in recovery and resolution associated with ultrathin as opposed to conventional PAGIF such as is encountered with t-RNA (127) and the various proteins studied (70). These losses are of concern to those wishing to retain antibody immunological reactivity during preparative procedures.

Such losses are not restricted to AK.Monooxygenases as a

class have seldom been electrofocused. A few rare studies attempting to electrofocus tryptophan (TP-OH) (122) and tyrosine hydroxylase (T-OH) (123) have been reported, but no details were available until success was reported with T-OH (121). A prohibitive loss of rabbit hindbrain TP-OH activity is shown (122) when the protein is focused in 440 ml columns and it is implied that the activity peak which emerges is not well resolved since the pI could only be presented as being "between 5.2 and 5.7". Details concerning T-OH were equally scanty until an excellent work on the subject was published (121). A reported success at electrofocusing T-OH from the bovine adrenal gland has been presented (123), but no detailed information is available.

In contrast to published results (122), the author has recently electrofocused the same enzyme from the rabbit hindbrain in 20 ml minicolumns with 45%-55% recovery and a sharp pI at 5.1, just outside the range which had been reported (11). One should also note that the scale of the column procedure is always much greater than that of the PAGIF procedure and, thus, may be a contributing factor to any spectral differences encountered across procedures. The inability to electrofocus either T-OH (excluding a brief report, 123), or TP-OH in conventional columns whereas they can be focused in minicolumns, illustrates that size dimensions or other factors related to the scale of a procedure are not to be underrated. Little is known as to why this should be true for both column and gel procedures.

Minicolumn electrofocusing may hold a great deal of promise for the detection or the large scale purification of immunologically important proteins, particularly IgA, IgM, and some immune complexes, that is to say proteins or protein complexes which do not readily enter gels (17, 50, 51, 98). This technique is superior to preparative procedures on agarose gels because mg quantities of a single protein can be isolated. It is superior to purification on conventional 110 ml and 440 ml columns because there is less chance of losing protein activity or immunological reactivity. Minicolumns are available in 20 ml and 30 ml sizes from International Biological Products (6065 Mission Gorge Road, Suite 107, San Diego, CA 92120). A 50 ml column is also being tested and a 10 ml microcolumn has been designed which may be used in microscale electrofocusing procedures (105, 128).

E) Gel standardization

Many factors dictate that the amount of protein applied to PAGIF gels should be standardized. This problem is not unique to immunological procedures but at least two factors render this an issue of greater concern in immunology than elsewhere. IgG is synthesized intrathecally by MS patients, causing CSF concentrations to differ in different patients and during exacerbations and remissions (129, 130). Thus, standardized protein concentrations are required for the accurate analysis of disease and reproducible analytical results. Even normal CSF and serum samples show several alkaline PAGIF bands when large amounts of IgG are applied (131). The amount of IgG thought to be optimal varies greatly from 5 micrograms (50) to 30 micrograms (101). Too little protein results in faint bands and gel patterns that cannot be accurately assessed (50). The amount of protein applied to gels can be determined by any standard procedure of protein determination such as the Lowry (132) or Bradford (133) methods. The applied protein itself contributes to gel artifacts. Gel regions of high protein concentration tend to form a small pH plateau due to the buffering capacity around its pI. This can affect zone sharpness (109).

As with any gel procedure, too little protein gives faint bands whereas too much can cause bands to smear or adjacent bands to fuse. Other factors related to visualizing protein bands are not directly related to the amount of protein applied to the gel. These problems can be reduced or eliminated if other factors besides the amount of sample applied are included among one's considerations. For example, ultrathin gels can show two protein bands where only a single fused band is apparent on conventional gels (70, 76). Different staining techniques (56, 57) and different methods of fixation (51, 57) also contribute to artifactual gel bands.

III. Analysis and Comparison of Spectrotypes

A) Value in disease detection and analysis

Isoelectric focusing has proved to be a valuable clinical tool for disease detection and analysis. Although its potential value to detect a number of diseases has been tested (100), it has proved most promising in the study of immunological disorders (50, 100-104), particularly multiple sclerosis (MS) (50, 101, 103), and gene disorders (16, 66-68, 135). It may have potential value in diagnosing cystic fibrosis (135, 136, 137, 168). It has been found (100) that PAGIF was of value in studying the CSF protein changes associated with a series of neurological disorders, including chronic and acute idiopathic polyneuropathy, myelopathy, meningo-encephalitis, sclerosing panencephalitis (SSPE), brain atrophy of various etiologies, intracranial tumours, brain tumours, acute cerebrovascular disorders and multiple sclerosis. Brief summaries of some of the data are presented in Tables 1 and 2. Most of these disorders are characterized by abnormal protein fractions with pI values at 5.9. MS and polyneuropathy are characterized by abnormal fractions with alkaline pI values between 6.5 and 9.5, but fractions with pI values between 8.3 and 9.5 are of particular note. Abnormal CSF-IgG fractions with pI values between 8.5 and 9.5 in MS patients have been reported in a number of

Table 1. Disorders with abnormal IgG fractions detected by PAGIF

Clinical Disorders	% cases with abnormal Ig G fractions			% fractions with p.I's between 8.5-9.5	Reference number
	serum	CSF	both		
1.) acute idiopathic polyneuropathy	-	15%	40%	82-83%	100
2.) chronic and recurrent idiopathic polyneuropathy	-	3	20%	63%	100
3.) sequelar of acute idiopathic polyneuropathy	-	-	46%	-	100
4.) multiple sclerosis					
a.)	-	95%	-	-	100
b.)	-	90%	-	10%+	103
c.)	9%	81%	-	81%*	101
d.)	varies, see report			-#	50

\+ pI range 8.5 to 9.3
* pI range 8.1 to 9.3
\# pI range 6.5 to 9.5

Disorders with abnormal IgG fractions detected by PAGIF.

studies (50, 100, 101, 103, 104). Unfortunately, many of the problems discussed so far are associated with proteins having alkaline pI values. In particular, PAGIF loses much of its accuracy when cationic proteins (23, 41, 108) or large proteins (17, 50, 51, 108) are studied because of 1) cathodic drift, 2) electroendosmosis, and 3) size restrictions due to limitations in the gel pore size. This part of the review will draw attention to some of these problems as they may be encountered in clinical literature and suggest recent techniques that can be usefully applied to overcome them.

Table 2. Disorders in which abnormal IgG fractions were found with PAGIF

Clinical Disorders	% cases with abnormal IgG fraction: serum	CSF	both	pI of abnormal fraction	Reference number
1.) Myelopathy	-	24%	-	5.9	100
2.) a.) Meningo-encephalitis & sclerosing panencephalitis (SSPE) (acute)	-	10%	-	5.9	100
b.) sequalae	-	40%	-	5.9	100
3.) Brain atrophy of various etiologies	-	33-6%*		5.9	100
4.) Intracranial tumors	-	38%		5.9	100
5.) Acute cerebrovascular disorders	-	20%	-	5.9	100
6.) Brain tumors	-	45%	-	5.9	100
7.) multiple sclerosis					
a.)	-	60.5%	-	6.5-8.5	100
b.)	9%	81%	-	8.1-9.3	101
c.)	-	90%	-	7.8-9.3	103
d.)	varies, see report			3 ranges, see report	50
8.) Polyneuropathy					
a.)		82-83%		6.5-8.5	100
b.)		63%		8.5-9.5	

*-a number of cases with presenile dementia and senile dementia had abnormal IgG fractions in CSF only or in both serum and CSF, but the actual percentage was not reported.

Disorders in which abnormal IgG fractions were found with PAGIF.

B) Analysis of body fluids

Isoelectric focusing has been of widespread value as a means of analyzing blood, serum, and CSF. PAGIF is a superior tool to gel electrophoresis to study blood, serum and CSF (101, 103, 138, 139, 169). Today the procedure is widely used for blood analysis in clinical situations (106, 140) and forensic science (39, 40, 141) as well as to detect signs of abnormal immune responses in a number of neurological (100) and genetic (26, 66-68, 135) disorders and to detect gene markers (135). Our focus will be on its neuroimmunological value.

From an immunological point of view, the CNS is a separate

entity from the rest of the immune system. Many CNS immune reactions occur independent of those outside the nervous system (142). This separation, which is valuable for diagnosing immunological disorders originating within the CNS, is facilitated by the blood-brain barrier which restricts the passage of unique CNS immune proteins from the blood of the brain and from CSF (143). Although the continuous cerebrovascular endothelium, epithelia of the choroid plexus, and arachnoid and perineurium restrict the passage of most proteins out of the nervous system, limited transfer is possible through nonbarrier regions and the choroid plexus, as well as transport by pinocytosis through arteriolar endothelium. The size and shape of proteins, rather than the molecular weight, are critical for protein filtration at the blood-brain barrier (144). Large proteins such as IgM (MW about 900,000) usually cannot cross the blood-brain barrier.

Restricted passage of proteins across the blood-brain barrier results in a differential distribution of oligoclonal IgG between the blood and CSF in some neurological disorders. Of most interest is the distribution of oligoclonal IgG's between the serum and CSF of MS patients. Oligoclonal CSF immunoglobulins are only rarely found in neurological disorders other than MS and infectious CNS diseases (142). Many investigators postulate that the oligoclonal CSF-IgG bands present in MS patients indicate the synthesis of specific antibodies within the nervous system (50, 142) and PAGIF data has contributed to this conclusion. The CSF of MS patients often reveals IgG bands not present in normal CSF (50, 101). Additionally, these abnormal CSF bands are not present in the serum of the same individual, or else exist in a greatly reduced serum concentration, i.e. 5-10 fold less in the serum than in the CSF (50). Such increases in the CSF protein concentration have been recognized since the early reported findings (145).

Recent technical improvements may eliminate PAGIF artifacts and thus prove valuable to the clinician and clinical laboratory analyst. Many of the IgG bands unique to the CSF of MS patients have highly alkaline pI values (50, 100, 101) but technical factors other than disease can contribute to the number of alkaline bands. One study (101) notes that the application of too much IgG sample to the gel or the extended storage of CSF each contribute to the number of alkaline PAGIF bands. Regarding the second point, it is observed (101) that storing CSF samples for more than one year can alter the pI from 8.0 to 9.3, probably due to protein denaturation.

At least three recent PAGIF modifications discussed in part II may prove valuable for the accurate clinical analysis of proteins with alkaline pI values. Briefly, these include 1) the use of ultrathin gels (70, 76), 2) the use of a two-stage PAGIF method (91), and 3) the 2-D procedure with IEF/SDS polyacrylamide gels (24, 76), particularly with ultrathin gels (76). Special attention to staining procedures, the choice of stain and of gel thickness, and the application of a standardized amount of IgG sample are also critical factors.

C) Special problems associated with IgA and IgM analysis

Analytical PAGIF studies are essentially restricted to detecting IgG since IgA and IgM are often too large to penetrate the gel matrix (50, 51, 147). Molecules with molecular weights exceeding 200,000 result in streaking due to the molecular sieving effects of the gel (98). Some studies have successfully analyzed body fluids for the presence of IgA by PAGIF without any reported difficulty (50, 148, 149), but these studies do not appear to have determined the MW of the IgA detected. Since IgA immunoglobulins vary from MW 180,000 to 500,000 those at the lower end of this range may have been focused while larger IgA molecules went undetected. The composite gel procedure which combines polyacrylamide and agarose gels (98) and a recent modification of isoelectric focusing called chromatofocusing (105) may eliminate some of the inability to detect IgA and IgM by electrofocusing.

Studies reveal that an agarose IEF system will allow for the resolution of large molecules such as IgM immunoglobulins

(150, 151). When the present author and coworkers (98) compared the agarose (0.8% agarose) and polyacrylamide (5% acrylamide) gels with a composite gel system consisting of 1.0% acrylamide plus 0.75% agarose for the study of IgA and IgM immunoglobulins they found the composite gel to be superior to either of the individual gels alone. Streaking observed in polyacrylamide gels was eliminated by switching to 0.8% agarose but the resultant spectral pattern was not well resolved. To illustrate their point the abovementioned workers (150, 151) present photos of the IgM spectrotype obtained from the agarose and composite gels and remark that the IgM spectrotypes focused as "two poorly resolved bands" (pI 5.8-6.3) on the agarose (although the present author could only discern a single poorly defined band), whereas the composite gel yielded multiple, clearly discernible bands (pI 5.6-6.4). The improved resolution and multiple bands cannot be attributed to the dissociation of the large molecules since dissociated heavy and light chains give markedly different spectrotypes from the parent molecule (153). Although the composite system permits for the rapid, reproducible, high resolution of intact polymeric IgA and IgM immunoglobulins, the IgA's and IgM's did not enter any of the gel systems studied when loaded at the cathode. Thus, IgA and IgM, but not IgG, must be loaded at the anode. A similar composite gel system which combines agarose and starch gels (154) may also prove valuable to analyze IgA and IgM immunoglobulins.

IgA and IgM immunoglobulins can also be resolved by the microscale sucrose system described (105, 128, 155). This system requires a sensitive radioimmunoassay to detect the focused components, but is rather slow and tedious. The microscale sucrose gradient system yields results comparable to those obtained with the composite gel system (98). The minicolumns used in a recently reported glycerol density gradient technique (22) can be replaced by 10 ml microcolumns (available from International Biological Products, San Diego) and should be as valuable as the microscale sucrose system although it may be subject to the same limitations. Glycerol can be replaced by sucrose or ethylene glycol (1). The high resolution of gel systems may not be maintained, but microscale density gradient systems can be combined with radioimmunoassay so as to detect newly formed IgA or IgM immunoglobulins lost on PAGIF gels. If certain immunoglobulins are uniquely present in a disease state, then high resolution is not a concern as long as differences between the two states can be detected. Microscale procedures with microcolumns should provide high resolution and reproducibility, but adjacent bands will remix somewhat during elution from the column. New gel matrices may also eliminate the problems associated with detecting IgA's and IgM's by PAGIF. One study (156) describes a new, highly cross-linked N,N'-(1,2-dihydroxyethylene) bisacrylamide gel which is hydrophilic, highly porous, and can be conveniently used for either electrophoresis or isoelectric focusing.

D) Antigen-antibody complexes

One can encounter the same problem with antigen-antibody complexes as with IgA and IgM immunoglobulins, that is, the complex may not enter the gel. Theoretical concepts relating to macromolecular isomerizations, protein-ligand interactions, and association-dissociation reactions as developed (6-8, 22, 146) are of particular value for interpreting IEF data related to antigen-antibody complexes. Antigens too large to enter the PAGIF gel matrix have been used (51) to demonstrate that thyroglobulin (Tg) antibodies are present on the surface of the PAGIF gel. A similar observation has been reported for immune complexes (17). When studying antigen-antibody complexes it is necessary to isolate the complex, dissociate it into its antigen and antibody components, and recombine the dissociated complex in order to confirm the presence of complexes. This allows for the distinction between immune complexes and other substances such as nonspecific IgG aggregates (157). Analytical and preparative isoelectric focusing has been used (104) to study the complex formed between soluble ribonucleoprotein nuclear protein (RNP) and the Sm antigen present in extracts of rabbit thymus acetone powder. Analytical PAGIF revealed two bands each at pH 5.2 and pH 5.4, for a total of four bands, in the thymus powder extract purified by ion-exchange chromatography. After digestion by ribonuclease, these bands disappeared whereas a faint new band focused at pH 6.1. This study used IEF and SDS separately but they can be combined into a single, 2-D procedure (24, 76). The success in detecting immune complexes is consistent with the hypothesis (6, 7) that interaction between two or more different macromolecules can alter the net charge and IEF profile of each macromolecule. The same idea is expressed elsewhere (104).

In another study (85), different sera were mixed and mixed and unmixed sera focused contiguously on agarose to determine the alignment of IgG-Frag A spectrotypes (ST) among different donors. A second set of mixing experiments was performed with sera of donors thought to have repeat IgG-Frag A ST as determined by close focusing analysis. Repeat IgG-Frag A ST were detected by mixing sera from two donors before IEF and comparing the mixed and unmixed sera from the same two donors. These experiments demonstrate that IEF can be used to detect the existence of shared complexes between different donors formed in response to immunization. Various studies show that immunization can cause the appearance of new antibodies and immune complexes in the circulating blood and serum (62, 85, 153, 158-161, 169).

A relatively new technique known as chromatofocusing shows considerable promise for studying immune complexes (105). Chromatofocusing is a modified version of isoelectric focusing in which proteins bound to an ion exchange resin in the order of their pI values are eluted with a pH gradient (162, 163). It can be performed in 8 M urea, thus allowing the dissociation of immune complexes. Similarly, by dissociating immune complexes using acidic pH in a one-step preparative procedure, it has been possible to study immune complexes associated with breast cancer (17, 164, 165). Chromatofocusing was used (84) to develop a sensitive method to analyze circulating IgG immune complexes without detailed knowledge as to the nature of the antigen or the specificity of the antibodies involved. Soluble BSA:anti-BSA were used as model immune complexes and isolated from the serum. BSA and anti-BSA were clearly separated by chromatofocusing in 8 M urea resulting in a high purity anti-BSA species which retained its immunological function during fractionation.

In a similar study of the same complex using PAGIF (17), the isoelectric focusing procedure itself was found to separate antigen and antibody according to their pI values. The recovered antigen and antibody are highly purified and retain their immunological reactivity. Thus, either chromatofocusing or PAGIF are useful tools to study immune complexes.

IV. Concluding Remarks

The PAGIF procedure has brought a fundamental change to the value of analytical gels in immunology. This change has been aided by the relatively new theoretical extensions described (6-11) which have included the interaction of proteins into IEF theory. PAGIF can be combined with other techniques such as SDS gel electrophoresis, ELISA and protein blotting so as to increase the value and usefulness of the combined techniques. The problem with PAGIF is that many artifacts are associated with protein bands located in the alkaline region of the gel, the region of most interest to the analytical immunologist. Several diverse techniques have been developed to overcome this problem, but one must still proceed with caution and be intimately familiar with the causes of these artifacts when using PAGIF as an analytical tool. Even with the numerous artifacts, PAGIF is the most valuable form of electrofocusing in immunology. Other focusing techniques may soon prove of great value also, particularly composite gel procedures and chromatofocusing. Isoelectric focusing procedures will remain an integral part of immunological techniques for years to come.

Acknowledgements

I wish to express my gratitude to Dr. Marion Hierowski and Dr. Robert Roskoski, Jr. for their critical review of and comments on this manuscript, Mrs. Linda Armstrong for secretarial assistance, and Mrs. Josephine Roussell for providing material important for the preparation of the manuscript. This work was supported by NS-15994 from the U.S. Public Health Service.

References

1. Vesterberg, O.: Isoelectric focusing of proteins. Meth. Enzymol. 22, 389-411 (1971).

2. Vesterberg, O.: Isoelectric focusing of proteins in polyacrylamide gels. Biochim. biophys. Acta 257, 11-19 (1972).

3. Righetti, P.G., Gianazza, E., Ek, K.: New developments in isoelectric focusing. J. Chromatog. 184, 415-456 (1980).

4. Allen, R.C.: Applications of polyacrylamide gel electrophoresis and polyacrylamide gel isoelectric focusing in clinical chemistry. J. Chromatog. 146, 1-32 (1978).

5. Chrambach, A.: Electrophoresis and electrofocusing on polyacrylamide gel in the study of native macromolecules. Mol. Cell. Biochem. 29, 23-46 (1980).

6. Cann, J.R.: Multiband isoelectric focusing patterns produced by macromolecular interactions. Meth. Enzymol. 61, 142-147 (1979).

7. Cann, J.R., Gardiner, K.J.: Isoelectric focusing of interacting systems. IV. Interaction of macromolecules with each other and with ligands. Biophys. Chem. 10, 203-210 (1979).

8. Cann, J.R., Stimpson, D.I., Cox, D.J.: Isoelectric focusing of interacting systems. III. Carrier ampholyte-induced macromolecular association or dissociation into subunits. Anal. Biochem. 86, 34-49 (1978).

9. Cann, J.R., Stimpson, D.I.: Isoelectric focusing on interacting systems. I. Carrier ampholyte-induced macromolecular isomerization. Biophys. Chem. 7, 103-114 (1977).

10. Stimpson, D.I., Cann, J.R.: Isoelectric focusing of interacting systems. II. pH-Dependent conformational transitions. Biophys. Chem. 7, 115-119 (1977).

11. Jackson, J.H., Schott, W., Knapp, S., Mandell, A.J.: Isoelectric focusing on minicolumns: Preliminary study of tryptophan hydroxylase from the rabbit hindbrain. (Submitted for publication).

12. Williams, R.J.P.: The conformational properties of proteins in solution. Biol. Rev. 54, 389-437 (1979).

13. Constans, J., Kühnl, P., Viau, M., Spielmann, W.: A new procedure for the determination of Transferrin^{C} (Tf^{C}) subtypes by isoelectric focusing. Existence of two additional alleles, Tf^{C4} and Tf^{C5}. Hum. Genet. 55, 111-114 (1980).

14. Giuliani, A., Hassan, H.J., Casalbore, P., Marini, L., Orlando, M., Tentori, L.: Structural or functional heterogeneity of normal human serum albumin, alloalbumin, bisalbumin. Clin. chim. Acta. 113, 43-49 (1981).

15. Hořejší, V., Tichá, M.: Affinity isoelectric focusing in polyacrylamide gel - a method to detect ligand-binding proteins. Anal. Biochem. 116, 22-26 (1981).

16. Fernö, M., Borg, A., Norgren, A.: A comparison of two steroid receptor assays in breast cancer: Dextran coated charcoal and isoelectric focusing. Anticancer Res. 3, 243-246 (1983).

17. Maidment, B.W. Jr., Papsidero, L.D., Chu, T.M.: Isoelectric focusing - a new approach to the study of immune complexes. J. Immunol. Meth. 35, 297-306 (1980).

18. Sogami, M., Petersen, H.A., Foster, J.F.: The microheterogeneity of plasma albumins. V. Permutations in disulfide pairings as a probable source of microheterogeneity in bovine albumin. Biochemistry 8, 49-58 (1969).

19. Wrange, Ö.: Isoelectric focusing of steroid hormone receptors in slabs of polyacrylamide gel. Breast Cancer Res. Treat. 3, 97-102 (1983).

20. Righetti, P.G., Brown, R.P., Stone, A.L.: Aggregation of heparin and other acidic polysaccharides in isoelectric focusing. Biochim. biophys. Acta 542, 232-244 (1978).

21. Kaplan, L.J., Foster, J.F.: Isoelectric focusing behaviour of bovine plasma albumin, mercaptalbumin, and beta-lactoglobulins A and B. Biochemistry 10, 630-636 (1971).

22. Jackson, J.H., Russell, P.J.: Characteristics of the isoelectric focusing procedure: Importance of column size, pH, and protein-protein interactions. Anal. Biochem. 137, 41-49 (1984).

23. Valkonen, K.H., Piha, R.S.: Isoelectric focusing and isoelectric points of bovine liver histones. Anal. Biochem. 104, 499-505 (1980).

24. Scheele, G., Pash, J., Bieger, W.: Identification of proteins according to biological activity following separation by two-dimensional isoelectric focusing/sodium dodecyl sul-

fate gel electrophoresis: Analysis of human exocrine pancreatic proteins. Anal. Biochem. 112, 304-313 (1981).

25. Klose, J.: Isoelectric focusing and electrophoresis combined as a method for defining new point mutations in the mouse. Genetics 92, s13-s24 (1979).

26. Hoste, B.: Group-specific component (Gc) and Transferrin (Tf) subtypes ascertained by isoelectric focusing. Hum. Genet. 50, 75-79 (1979).

27. Gjerde, J.: Thin layer iso-electro focusing of intracellular bacterial protein in classification of some representatives of the families Enterobacteriaceae and Vibrionaceae. Acta Path. Microbiol. Immunol. Scand. Sect. B 90, 383-388 (1982).

28. Saravis, C.A., O'Brien, M., Zamcheck, N.: Direct tissue isoelectric focusing in agarose. J. Immunol. Meth. 29, 97-100 (1979).

29. Saravis, C.A., Zamcheck, N.: Isoelectric focusing in agarose. J. Immuno. Meth. 29, 91-96 (1979).

30. Allan, B.J., White, T.T., Kirk, J., Schilling, J.J.: Techniques for reproducible transient-state isoelectric focusing of human pancreatic secretory proteins with computer-assisted pattern matching, averaging, and analysis. Anal. Biochem. 113, 1-12 (1981).

31. Kopelovich, L., Wolfe, G., Pfeffer, L.: Segregation of histone fractions from purified rat pancreas nuclei by isoelectric focusing. Proc. Soc. exp. Biol. Med. 151, 400-406 (1976).

32. Alhadeff, J.A., Cimino, G., Janowsky, A., O'Brien, J.S.: Human liver glycoprotein sialytransferase. Biochim. biophys. Acta 484, 307-321 (1977).

33. Moore, H.D.M., Hibbit, K.G.: Isoelectric focusing of boar spermatozoa. J. Reprod. Fertil. 44, 329-332 (1975).

34. Haff, L.A., Lasky, M., Manrique, A.: A new technique for desorbing substances tightly bound to affinity gels: Flatbed electrophoretic desorption in sephadex via isoelectric focusing (FEDS-IEF). J. Biochem. Biophys. Meth. 1, 275-286 (1979).

35. McGuire, J.K., Miller, T.Y., Tipps, R.W., Snyder, R.S., Righetti, P.G.: New experimental approaches to the isoelectric fractionation of cells. J. Chromatog. 194, 323-333 (1980).

36. Sherbet, G.V., Lakshmi, M.S., Rao, K.V.: Characterisation of ionogenic groups and estimation of the net negative electric charge on the surface of cells using natural pH gradients. Exp. Cell Res. 70, 113-123 (1972).

37. Sherbet, G.V., Lakshmi, M.S.: Characterisation of Escherichia coli cell surface by isoelectric equilibrium analysis. Biochim. biophys. Acta 298, 50-58 (1973).

38. Wykle, B., Gillett, T.A., Addis, P.B.: Myoglobin heterogeneity in pigs with PSE and normal muscle by an improved isoelectric focusing technique. J. Animal Sci. 47, 1260-1264 (1978).

39. Slattery, W.J., Sinclair, A.J.: Differentiation of meat according to species by the electrophoretic separation of muscle lactate dehydrogenase and esterase isoenzymes and isoelectric focusing of soluble muscle proteins. Aust. vet. J. 60, 47-51 (1983).

40. Oshima, M., Inoue, T., Hara, M.: Identification of species specific hemoglobin by isoelectric focusing. Forensic Sci. Int. 20, 277-286 (1982).

41. Delincée, H., Radola, B.J.: Determination of isoelectric points in thin-layer isoelectric focusing: The importance of attaining the steady state and the role of CO_2 interference. Anal. Biochem. 90, 609-623 (1978).

42. Radola, B.J.: Thin-layer isoelectric focusing of proteins. Biochim. biophys. Acta 194, 335-338 (1969).

43. Delincée, H., Radola, B.J.: Thin-layer isoelectric focusing on Sephadex layers of horseradish peroxidase. Biochim. biophys. Acta 200, 404-407 (1970).

44. Frater, R.: Artifacts in isoelectric focusing. J. Chromatog. 50, 469-474 (1970).

45. Salaman, M.R., Williamson, A.R.: Isoelectric focusing of proteins in the native and denatured states. Anomalous behaviour of plasma albumin. Biochem. J. 122, 93-99 (1971).

46. Beeley, J.A., Stevenson, S.M., Beeley, J.G.: Polyacrylamide gel isoelectric focusing of proteins: Determination of isoelectric points using an antimony electrode. Biochim. biophys. Acta 285, 293-300 (1972).

47. Radola, B.J.: Isoelectric focusing in layers of granulated gels. I. Thin-layer isoelectric focusing of proteins. Biochim. biophys. Acta 295, 412-428 (1973).

48. Bours, J.: The use of protein markers in thin-layer isoelectric focusing. Sci. Tools 20, 29-34 (1973).

49. Gianazza, E., Righetti, P.G., Bordi, S., Papeschi, G.: pH Determinations in isoelectric focusing with an iridium electrode. In: "Electrofocusing and Isotachophoresis", Eds. Radola, B.J., Graesslin, D., Walter de Gruyter, Berlin, New York, pp. 173-179 (1977).

50. Laurenzi, M.A., Mavra, M., Kam-Hansen, S., Link, H.: Oligoclonal IgG and free light chains in multiple sclerosis demonstrated by thin-layer polyacrylamide gel isoelectric focusing and immunofixation. Ann. Neurol. 8, 241-247 (1980).

51. Nye, L., Roitt, I.M.: Isoelectric focusing of human antibodies directed against a high molecular weight antigen. J. Immunol. Meth. 35, 97-103 (1980).

52. Danno, G.I.: Isoelectric focusing of proteins separated by SDS-polyacrylamide gel electrophoresis. Anal. Biochem. 83, 189-193 (1977).

53. Zechel, K.: On the resolution of polypeptides by isoelectric focusing in polyacrylamide gels in the presence of urea and nonidet-p40. Anal. Biochem. 83, 24-51 (1977).

54. Reisner, A.H., Nemes, P., Bucholtz, C.: The use of Coomassie Brilliant Blue G 250 perchloric acid solution for staining in electrophoresis and isoelectric focusing on polyacrylamide gels. Anal. Biochem. 64, 509-516 (1975).

55. Vesterberg, O., Hansen, L., Sjosten, A.: Staining of proteins after isoelectric focusing in gels by new procedures. Biochim. biophys. Acta 491, 160-166 (1977).

56. Reinhart, M.P., Malamud, D.: Protein transfer from isoelectric focusing gels: The native blot. Anal. Biochem. 123, 229-235 (1982).

57. Allen, R.E., Masak, K.C., McAllister, P.K.: Staining protein in isoelectric focusing gels with fast green. Anal. Biochem. 104, 494-498 (1980).

58. Kaplow, L.S.: Substitute for benzidine in myeloperoxidase stains. Am. J. Clin. Path. 63, 451 (1975).

59. Walker, R.W.H., Keir, G., Johnson, M.H., Thompson, E.J.: A rapid method for detecting oligoclonal IgG in unconcentrated CSF, by agarose isoelectric focusing, transfer to cellulose nitrate and immunoperoxidase staining. J. Neuroimmunol. 4, 141-148 (1983).

60. Culbertson, C.G.: Soil ameba infection. Specific indirect immunoenzymatic (peroxidase) staining of formalin-fixed paraffin sections. Am. J. Clin. Path. 63, 475-482 (1975).

61. Williamson, A.R.: Antibody isoelectric spectra. Analysis of the heterogeneity of antibody molecules in serum by isoelectric focusing in gel and specific detection with hapten. Europ. J. Immunology 1, 390-394 (1971).

62. Keck, K., Grossberg, A.L., Pressman, D.: Specific characterization of isoelectrofocused immunoglobulins in polyacrylamide gel by reaction with ^{125}I-labelled protein antigens or antibodies. Europ. J. Immunology 3, 99-102 (1973).

63. Nicolotti, R.A., Briles, D.E., Schroer, J.A., Davie, J.M.: Isoelectric focusing of immunoglobulins: Improved methodology. J. Immunol. Meth. 33, 101-115 (1980).

64. Cooke, A., North, J., Feinstein, A.: Antibody of restricted heterogeneity secreted by spleen fragments. Nature New Biol. 240, 152-154 (1972).

65. Martin, M.J., Landon, J.: The use of ammonium sulfate and dextran together with second antibody at the separation stage in radioimmunoassay. In: "Radioimmunoassay in Clinical

Biochemistry", Ed. Pasternak, C.A., Heydon and Son Ltd., London, pp. 269-281 (1975).

66. Massi, G., Fabiano, A., Ragusa, D., Auconi, P.: Characterization of alpha-1-antitrypsin by isoelectric focusing on an ultrathin polyacrylamide gel layer. Hum. Genet. 53, 91-95 (1979).

67. Pascali, V.L., Ranalletta, D., Auconi, P.: Improved typing of human serum transferrin by isoelectric focusing on ultrathin layer polyacrylamide slab gels. Hum. Genet. 61, 39-41 (1982).

68. Cortivo, P., Biasiolo, M., Crestani, C., Scorretti, C., Benciolini, P.: The polymorphism of transferrin by ultrathin-layer isoelectric focusing. Tf Phenotypes and subtypes in the population of Padua. For. Sci. Int. 24, 65-68 (1984).

69. Bours, J.: Isoelectric focusing of lens crystallins in thin-layer polyacrylamide gels. A method for detection of soluble proteins in eye lens extracts. J. Chromatog. 60, 225-233 (1971).

70. Görg, A., Postel, W., Westermeier, R.: Ultrathin-layer focusing in polyacrylamide gels on cellophane. Anal. Biochem. 89, 60-70 (1978).

71. Righetti, P.G., Drysdale, J.: Analytical IEF. In: "Isoelectric Focusing", Eds. Work, T.S., Work, E., North-Holland, Amsterdam/Oxford; American Elsevier, New York, pp. 419-500 (1976).

72. Görg, A., Postel, W., Westermeier, R.: A simple method for preparing precast polyacrylamide gel plates for isoelectric focusing. Z. Lebensmittelunters.u.-Forsch. 164, 25-28 (1977).

73. Soderholm, J., Allestam, P., Wadstrom, T.: A rapid method for isoelectric focusing in polyacrylamide gel. FEBS Lett. 24, 89-92 (1972).

74. Righetti, P.G., Righetti, A.B.B.: High voltage analytical and preparative isoelectric focusing. In: "Isoelectric Focusing", Eds. Arbuthnott, J.P., Beeley, J.A., Butterworths, London, pp. 114-131 (1975).

75. Haglund, H.: Properties of ampholine. In: "Isoelectric Focusing", Eds. Arbuthnott, J.P., Beeley, J.A., Butterworths, London, pp. 3-23 (1975).

76. Görg, A., Postel, W., Westermeier, R., Giananzza, E., Righetti, P.G.: Gel gradient electrophoresis, isoelectric focusing and two-dimensional techniques in horizontal, ultrathin polyacrylamide layers. J. Biochem. Biophys. Meth. 3, 273-284 (1980).

77. Vesterberg, O., Hansen, L.: Staining of proteins in polyacrylamide gels. In: "Electrofocusing and Isotachophoresis", Eds. Radola, B.J., Graesslin, D., Walter de Gruyter, Berlin, New York, pp. 123-133 (1977).

78. Scheele, G.: Two-dimensional gel analysis of soluble proteins. Characterization of guinea pig exocrine pancreatic proteins. J. biol. Chem. 250, 5375-5385 (1975).

79. O'Farrell, P.H.: High resolution two-dimensional electrophoresis of proteins. J. biol. Chem. 250, 4007-4021 (1975).

80. Wright, G.L. Jr., Farrell, K.B., Roberts, D.B.: An evaluation of gradient acrylamide gel electrophoresis and acrylamide gel isoelectric focusing for the primary separation of complex mixtures of proteins: Comparison of one- and two-dimensional analytical procedures. Biochim. biophys. Acta 295, 396-411 (1973).

81. Emes, A.V., Latner, A.L., Martin, J.A.: Electrofocusing followed by gradient electrophoresis: A two-dimensional polyacrylamide gel technique for the separation of proteins and its application to the immunoglobulins. Clin. chim. Acta 64, 69-78 (1975).

82. Engvall, E., Perlman, P.: Enzyme-linked immunosorbent assay. Elisa 3. Quantitation of specific antibodies by enzyme-labelled anti-immunoglobulin in antigen-coated tubes. J. Immunol. 109, 129-135 (1972).

83. Towbin, H., Staehlin, T., Gordon, J.: Electrophoretic transfer of proteins from polyacrylamide gels to nitrocellulose sheets: Procedure and some applications. Proc. Nat. Acad. Sci. 76, 4350-4354 (1979).

84. Kneba, M., Krieger, G., Kehn, A., Bause, I., Nagel, G.A.: Chromatofocusing combined with ELISA technique. A sensitive method for the analysis of immune complexes. J. Immunol. Meth. 61, 233-243 (1983).

85. Morrow, C.D., Dorey, F., Stevens, R.H.: Isoelectric focusing of human anti-diphtheria toxid antibodies: Identical spectrotypes of anti-fragment A antibodies with the same IgG subclass and light chain constant regions are expressed in multiple individuals. J. Immunol. 130, 818-823 (1983).

86. Gershoni, J.M., Palade, G.E.: Protein blotting: Principles and applications. Anal. Biochem. 131, 1-15 (1983).

87. Southern, E.M.: Detection of specific sequences among DNA fragments separated by gel electrophoresis. J. Mol. Biol. 98, 503-517 (1975).

88. Mandrell, R.E., Zollinger, W.D.: Use of a zwitterionic detergent for the restoration of the antibody-binding capacity of electroblotted meningoccal outer membrane proteins. J. Immunol. Meth. 67, 1-11 (1984).

89. Alwine, J.C., Kemp, D.J., Stark, G.R.: Method for detection of specific RNAs in agarose gels by transfer to diazobenzyloxymethyl-paper and hybridization with DNA probes. Proc. Nat. Acad. Sci. 74, 5350-5354 (1977).

90. Friedenson, B., Soong, C.-J.: A simple general method of exactly comparing different binding activities and antigenic

properties of antibodies by obtaining duplicate copies from a single isoelectric focusing gel. J. Immunol. Meth. 67, 235-242 (1984).

91. Thomas, J.M., Hodes, M.E.: Two-stage methods for alkaline-range analytical isoelectric focusing in thin-layer polyacrylamide gel: Application to cationic proteins in serum. Anal. Biochem. 90, 596-608 (1978).

92. Bjellqvist, B., Ek, K., Righetti, P.G., Gianazzi, E., Görg, A., Westermeier, R., Postel, W.: Isoelectric focusing in immobilized pH gradients: Principle, methodology and some applications. J. Biol. Biophys. Meth. 6, 317-339 (1982).

93. Dossi, G., Celentano, F., Gianazza, E., Righetti, P.G.: Isoelectric focusing in immobilized pH gradients: Generation of extended pH intervals. J. Biochem. Biophys. Meth. 7, 123-142 (1983).

94. Nguyen, N.Y., Chrambach, A.: Stabilization of pH gradients in buffer electrofocusing on polyacrylamide gel. Anal. Biochem. 82, 54-62 (1977a).

95. Nguyen, N.Y., Chrambach, A.: Stabilization of pH gradients formed by ampholine. Anal. Biochem. 82, 226-235 (1977b).

96. Svensson, H.: Isoelectric fractionation, analysis, and characterization of ampholytes in natural pH gradients. II. Buffering capacity and conductance of isoionic ampholytes. Acta chem. scand. 16, 456-466 (1962).

97. Righetti, P.G., Chrambach, A.: Hindered migration of amino acids toward their isoelectric positions in gel electrofocusing and facilitation of the migration at high ionic strength. Anal. Biochem. 90, 633-643 (1978).

98. Jackson, D.E., Skandera, C.A., Owen, J., Lally, E.T., Montgomery, P.C.: Isoelectric focusing of IgA and IgM in composite acrylamide-agarose gels. J. Immunol. Meth. 36, 315-324 (1980).

99. Dubart, A., Goossens, M., Beuzard, Y., Monplaisir, N., Testa, U., Basset, P., Rosa, J.: Prenatal diagnosis of hemoglobinopathies: Comparison of the results obtained by isoelectric focusing of hemoglobins and by chromatography of radioactive globin chains. Blood 56, 1092-1099 (1980).

100. Livrea, P., Zimatore, G.B., Simone, I.L., Lepore, V., Trojano, M., Loreto, M., de Blasi, A., Gennarini, G.F.: Isoelectric focusing and crossed immunoelectrofocusing of cerebrospinal fluid proteins in neurological disorders. Acta Neurol. 33, 501-517 (1978).

101. Olsson, J.E., Nilsson, K.: Gamma globulins of CSF and serum in multiple sclerosis: Isoelectric focusing on polyacrylamide gel and agar gel electrophoresis. Neurology 29, 1383-1391 (1979).

102. Beneditti, G.M., Taiuti, R., Bartolini, S., Amaducci, L.: L'uso dell' isoelettrofocusing nello studio delle proteine liquorali, sua applicazione nella diagnosi delle malattie del sistema nervoso. Riv. Pat. Nerv. Ment. 98, 347-362 (1977).

103. Mehta, P.D., Patrick, B.A., Wisniewski, H.M.: Isoelectric focusing and immunofixation of cerebrospinal fluid and serum in multiple sclerosis. J. Clin. Lab. Immunol. 6, 17-22 (1981).

104. Waelti, E.R., Hess, M.W.: Nuclear antigens: Characterization of acidic nuclear ribonucleoprotein-Sm-antigen complex by analytical and preparative isoelectric focusing. J. Immunol. Meth. 32, 177-184 (1980).

105. Vartdal, F., Bird, P.: Imprint immunofixation of IgG subclasses, using monoclonal subclass-specific antibodies. Scand. J. Immunol. 18, 367-370 (1983).

106. Lane, J.R., Bowles, K.J., Normansell, D.E.: Detection of oligoclonal IgG bands in cerebrospinal fluid by isoelectric focusing on thin layer agarose gels. Ann. Clin. Lab. Sci. 14, 146-150 (1984).

107. Gearhart, P.J., Sigal, N.H., Klinman, N.R.: Production of antibodies of identical idiotype but diverse immunoglobulin classes by cells derived from a single stimulated B cell. Proc. Nat. Acad. Sci. 72, 1707-1711 (1975).

108. Righetti, P.G., Tudor, G., Gianazza, E.: Effect of 2-mercaptoethanol on pH gradients in isoelectric focusing. J. Biochem. Biophys. Meth. 6, 219-227 (1982).

109. Låås, T., Olsson, I., Soderberg, L.: High voltage isoelectric focusing with pharmalyte: Field strength and temperature distribution, zone sharpening, isoelectric spectra, and pI determinations. Anal. Biochem. 101, 449-461 (1980).

110. Haglund, H.: Isoelectric focusing in pH gradients - a technique for fractionation and characterization of ampholytes. 19, 1-104 (1971).

111. Radola, B.J.: Isoelectric focusing in layers of granulated gels. II. Preparative isoelectric focusing. Biochim. biophys. Acta 386, 181-195 (1974).

112. McCormick, A.G., Wachslicht, H., Chrambach, A.: Preparative separation of hemoglobins A and S by gel electrofocusing, using selective zone elution by gel transposition between suitable anolytes and catholytes. Anal. Biochem. 85, 209-218 (1978).

113. Galante, E., Caravaggio, T., Righetti, P.G.: Binding of ampholine to transfer RNA. Biochim. biophys. Acta 442, 309-315 (1976).

114. Righetti, P.G., Gianazza, E.: Isoelectric focusing of

heparin. Evidence for complexing with carrier ampholytes. Biochim. biophys. Acta 532, 137-146 (1978).

115. Felgenhauer, K., Pak, S.J.: Detection of ampholine patterns. Ann. N.Y. Acad. Sci. 209, 147-153 (1973).

116. Bonitati, J.: Visualization of carrier ampholyte patterns in granular gel slabs. 2. Applications to preparative electrofocusing. J. Biochem. Biophys. Meth. 4, 49-69 (1981).

117. Yoshida, K.: A versatile apparatus for electrophoresis including isoelectrofocusing and its performance. Anal. Biochem. 129, 37-45 (1983).

118. Ebers, G.C., Rice, G.P., Armstrong, H.: Preparative isoelectric focusing in agarose. J. Immunol. Meth. 37, 315-323 (1980).

119. Prestidge, R.L., Kearn, M.T.W.: Preparative flatbed electrofocusing in granulated gels with natural pH gradients generated from simple buffers. Anal. Biochem. 97, 95-102 (1979).

120. Jackson, J.H., Schott, W., Russo, P.V., Mandell, A.J.: Isoelectric focusing on minicolumns: Comparison of tyrosine hydroxylase patterns from bovine and rat caudate. (Submitted for publication).

121. Meligeni, J., Weiner, N.: Isoelectric focusing of tyrosine hydroxylase. In: "Function and Regulation of Monoamine Enzymes: Basic and Clinical Aspects", Eds. Usdin, E., Weiner, N., Youdim, M.B.H., MacMillan Publishers Ltd., London, pp. 25-34 (1981).

122. Tong, J.H., Kaufman, S.: Tryptophan hydroxylase: Purification and some properties of the enzyme from rabbit hindbrain. J. biol. Chem. 250, 4152-4158 (1975).

123. Vigny, A., Henry, J-P.: Bovine adrenal tyrosine hydroxylase: Comparative study of native and proteolyzed enzyme, and their interaction with anions. J. Neurochem. 36, 483-489 (1981).

124. Russell, P.J. Jr., Horenstein, J.M., Goins, L., Jones, D., Laver, M.: Adenylate kinase in human tissues. I. Organ specificity of adenylate kinase isoenzymes. J. biol. Chem. 249, 1874-1879 (1974).

125. Flatmark, T.: Personal communication.

126. Abrams, S.: Personal communication.

127. Drysdale, J.W., Righetti, P.: Fractionation of nucleic acids by isoelectric focusing. Biochemistry 11, 4044-4052 (1972).

128. Montgomery, P.C., Skandera, C.A.: Molecular heterogeneity of secretory IgA anti-DNP antibodies. J. Immunol. 121, 111-114 (1978).

129. Olsson, J.E., Link, H., Muller, R.: Immunoglobulin abnormalities in multiple sclerosis. Relation to clinical parameters: Disability, duration, age of onset. J. Neurol. Sci. 27, 233-245 (1976).

130. Olsson, J.E., Link, H.: Immunoglobulin abnormalities in multiple sclerosis. Relation to clinical parameters: Exacerbations and remissions. Arch. Neurol. 28, 392-399 (1973).

131. Nilsson, K., Olsoon, J.E.: Analysis for cerebrospinal fluid proteins by isoelectric focusing on polyacrylamide gel: Methodological aspects of normal values, with special references to the alkaline region. Clin. Chem. 24, 1134-1139 (1978).

132. Lowry, O.H., Rosebrough, N.J., Farr, A.L., Randall, R.J.: Protein measurement with the folin phenol reagent. J. biol. Chem. 193, 265-275 (1951).

133. Bradford, M.: A rapid and sensitive method for the quantitation of microgram quantities of protein utilizing the principle of protein-dye binding. Anal. Biochem. 72, 248-254 (1976).

134. Fredriksson, S.: On the temperature dependence of isoelectric points of proteins with special reference to isoelectric focusing. In: "Electrofocusing and Isotachophoresis", Eds. Radola, B.J., Graesslin, D., Walter de Gruyter, Berlin, New York, pp. 71-83 (1977).

135. Nevin, G.B., Nevin, N.C., Redmond, A.O., Young, I.R., Tully, G.W.: Detection of cystic fibrosis homozygotes and heterozygotes by serum isoelectrofocusing. Hum. Genet. 56, 387-389 (1981).

136. Wilson, G.B., Fudenberg, H.H., Jahn, T.L.: Studies on cystic fibrosis using isoelectric focusing. I. An assay for detection of cystic fibrosis homozygote, and heterozygote carriers for serum. Pediatr. Res. 9, 635-640 (1975).

137. Wilson, G.B., Arnaud, P., Fudenberg, H.H.: Improved method for the detection of cystic fibrosis protein in the serum using the LKB Multiphor electrofocusing apparatus. Pediatr. Res. 12, 801-804 (1978).

138. Arnaud, P., Wilson, G.B., Koistinen, J., Fudenberg, H.H.: Immunofixation after electrofocusing: Improved method for specific detection of serum proteins with determination of isoelectric points. I. Immunofixation print technique for detection of alpha-1-protease inhibitor. J. Immunol. Meth. 16, 221-231 (1977).

139. Madaliński, A.W., Wilińska, L., Sagan, Z.: Isoelectric focusing of the human IgG light chains: A classification of fractions and their normal values. Immunobiology 165, 90-96 (1983).

140. Haimovich, J., Jaton, J.-C., Pink, J.R.L.: Studies on the heterogeneity and idiotypic specificity of anti-dinitrophenyl antibodies from inbred guinea pigs. Europ. J. Immunol. 4, 290-295 (1974).

141. Lawton, M.E., Sutton, J.G.: Species identification of deer blood by isoelectric focusing. J. For. Sci. Soc. 22, 361-366 (1982).

142. Link, H.: Some aspects of immune reactions of the brain. In: "Clinical Neuroimmunology", Ed. Rose, F.C., Blackwell Scientific Publications, Oxford, pp. 12-28 (1979).

143. Rapoport, S.I.: Permeability and osmotic properties of the blood-brain barrier. In: "Blood-Brain Barrier in Physiology Medicine", Raven Press, New York, pp. 87-127 (1976).

144. Felgenhauer, K.: Protein size and cerebrospinal fluid composition. Klin. Wschr. 52, 1158-1164 (1974).

145. Kabat, E.A., Moore, D.H., Landow, H.: An electrophoretic study of the protein components in cerebrospinal fluid and their relationship to the serum proteins. J. clin. Invest. 21, 571-577 (1942).

146. Jackson, J.H., Schott, W., Knapp, S., Mandell, A.J.: Isoelectric focusing on minicolumns used to characterize purified tryptophan hydroxylase from the rat midbrain. (Submitted for publication).

147. Bouman, H., Lüsebrink, W.D., Havstein, B., Stein, H.: Discrimination between monoclonal and heterogeneous immunoglobulin M by isoelectric focusing and radioautography. FEBS Lett. 64, 201-203 (1976).

148. Link, H., Laurenzi, M.A.: Immunoglobulin class and light chain type of oligoclonal bands in CSF in multiple sclerosis determined by agarose gel electrophoresis and immunofixation. Ann. Neurol. 6, 107-110 (1979).

149. Link, H., Muller, R.: Immunoglobulins in multiple sclerosis and infections of the nervous system. Arch. Neurol. 25, 326-344 (1971).

150. Rosen, A., Ek, K., Aman, P.: Agarose isoelectric focusing of native human immunoglobulin M and alpha 2-macroglobulin. J. Immunogenet. 28, 1-11 (1979).

151. Allsopp, B.A., Gibson, W.C.: Isoelectric focusing in agarose: A highly discriminatory method for the detection of enzyme heterogeneity. Ann. Trop. Med. Parasitol. 77, 169-173 (1983).

152. Vecoli, C., Prevost, F.E., Ververis, J.J., Medeiros, A.A., O'Leary, G.P.: Comparison of polyacrylamide and agarose gel thin-layer isoelectric focusing for the characterization of beta-lactamases. Antimicrob. Agents Chemother. 24, 186-189 (1983).

153. Hoffman, D.R., Grossberg, A.L., Pressman, D.: Anti-hapten antibodies of restricted heterogeneity: Studies on binding properties and component chains. J. Immunol. 108, 18-25 (1972).

154. Scott, A.C., Fowler, J.C.S.: Electrophoretic typing of glyoxalase I (GLO I) isoenzymes using a mixed starch/agarose gel. For. Sci. Int. 20, 287-294 (1982).

155. Press, J.L., Klinman, N.R.: Isoelectric analysis of neonatal monofocal antibody. Immunochemistry 10, 621-627 (1973).

156. Righetti, P.G., Brost, B.C.W., Snyder, R.S.: On the limiting pore size of hydrophilic gels for electrophoresis and isoelectric focusing. J. Biochem. Biophys. Meth. 4, 347-363 (1981).

157. Soltis, R.D., Wilson, I.D.: Are immune complexes immunoglobulin aggregates? In: "Protides of the Biological Fluids", Ed. Peeters, H., Pergamon Press, vol. 26, pp. 127-131 (1979).

158. Awdeh, Z.L., Williamson, A.R., Askonas, B.A.: Isoelectric focusing in polyacrylamide gel and its application to immunoglobulins. Nature 219, 66-67 (1968).

159. Awdeh, Z.L., Williamson, A.R., Askonas, B.A.: Heavy and light chains of a homogeneous immunoglobulin G. FEBS Lett. 5, 275-278 (1969).

160. Gerber, H., Barandum, S.: Isoelectric separation of urinary Bence-Jones proteins and light chains of human IgG. Immunochemistry 6, 469-471 (1969).

161. Vienne, A., Wikler, M.: Isoelectric focusing of proteolytic fragments from rabbit anti-streptococcal antibodies of restricted heterogeneity. FEBS Lett. 39, 160-163 (1974).

162. Sluyterman, L.A.E., Elgersma, O.: Chromatofocusing : Isoelectric focusing on ion-exchange columns. I. General principles. J. Chromatog. 150, 17-30 (1978).

163. Sluyterman, L.A.E., Wijdenes, J.: Chromatofocusing : Isoelectric focusing on ion-exchange columns. II. Experimental verification. J. Chromatog. 150, 31-44 (1978).

164. Maidment, B.W. Jr., Papsidero, L.D., Gamarra, M., Nemoto, T., Chu, T.M.: Isoelectric focusing analysis of soluble immune complexes bound to A-Sepharose. Anal. Biochem. 111, 336-342 (1981a).

165. Maidment, B.W. Jr., Papsidero, L.D., Nemoto, T., Chu, T.M.: Recovery of immunologically reactive antibodies and antigens from breast cancer immune complexes by preparative isoelectric focusing. Cancer Res. 41, 795-800 (1981b).

166. Russell, P.J., Booher, J., Orr, T., Saenz, R., Abrams, S.: Adenylate kinase isozymes of cat tissues. Comp. Biochem. Physiol. 67B, 605-611 (1980).

167. Gurney, M.E.: Suppression of sprouting at the neuromuscular junction by immune sera. Nature 307, 546-548 (1984).

168. Wilson, G.B., Fudenberg, H.H.: Studies on cystic fibrosis using isoelectric focusing. II. Demonstration of deficient proteolytic cleavage of alpha 2-macroglobulin in cystic fibrosis plasma. Pediatr. Res. 10, 87-96 (1976).

169) Chopin, N., Lacour, B., Thierry, D., Faure, A., Bailly, M.: Apport de l'isoélectrofocalisation à l'étude de l'hétérogénéité des immunoglobulines G chez l'homme avec ou sans symptomatologie proliférative. Ann. Biol. clin. 41, 99-104 (1983).

A NEW FLUOROIMMUNOASSAY OF BIOPTERIN AND NEOPTERIN IN HUMAN URINE

M. Sawada*, T. Yamaguchi*, T. Sugimoto[+], S. Matsuura[+], T. Nagatsu*§
*Laboratory of Cell Physiology, Department of Life Chemistry, Graduate School of Nagatsuta, Tokyo Institute of Technology, Yokohama 227, Japan
§Department of Biochemistry, Nagoya University School of Medicine, Nagoya 466, Japan
[+]Department of Chemistry, College of General Education, Nagoya University, Nagoya 464, Japan.

Introduction

Automated immunoassay is a relatively new and simplified assay procedure in clinical chemistry. An example is an immunoassay that works with homogeneous samples; the technique does not require the separation of bound from free unknown and is insensitive to background sources naturally present in the sample. Polarization fluoroimmunoassay (1, 2) in which a fluorescent molecule is substituted for a radioactive ligand used in radioimmunoassay, is suitable for this purpose since it does not need any separation steps. The other advantages of fluoroimmunoassay are avoidance of radioactivity, a much longer lifetime of the assay kit and less expensive equipment. In particular, recent reports (1, 3) have indicated the advantages of polarization fluoroimmunoassay. Polarization of fluorescence results from the relative orientation of the excitation and emission dipoles which is affected by changes in size, rotation restrictions, conformation of the fluorescent molecules, solvent viscosity and temperature (1). Polarization fluoroimmunoassay does not

§ Address for correspondence (T. Nagatsu).

Immunoassay Technology, Volume 1

suffer from decreased sensitivity due to the variation and magnitude of the sample's own fluorescence since it is based on measurement of changes in molecular size by antibody binding.

Aromatic amino acid monooxygenases, phenylalanine hydroxylase (4, 5), tyrosine hydroxylase (6, 7) and tryptophan hydroxylase (8-10) require a tetrahydropteridine as co-factor and play an important role in the biosynthesis of biogenic monoamines such as catecholamines (dopamine, norepinephrine and epinephrine) and indoleamines (serotonin and melatonin), which belong to the family of important neurotransmitters or hormones. L-erythro-Tetrahydrobiopterin is the natural co-factor for the three aromatic amino acid monooxygenases (5, 7, 8, 10).

Recent reports indicate that hyperphenylalaninemia, as a variant of phenylketonuria, is due to a deficiency of biopterin (11-13), and that the level of biopterin in the cerebrospinal fluid (14) and brain (15) of patients with Parkinson's disease is lower than that of controls. On the other hand, 7,8-dihydroneopterin triphosphate is an intermediate in the biosynthesis of tetrahydrobiopterin (16-19), and the level of neopterin in urine of patients with a deficiency of biopterin (23, 13) and of cancer patients (20, 21) has been reported to be higher than that of controls, Therefore, a simple and rapid assay of biopterin and neopterin in human urine is required for the screening of such patients. High-performance liquid chromatography with fluorometry (22) or with electrochemistry (23), bioassay of Crithidia fasciculata (24) and radioimmunoassay (25-30) have been used for this purpose. High-performance liquid chromatography is most frequently used; it is a sensitive technique and biopterin and neopterin can be measured simultaneously. However, it is rather time-consuming for screening many samples. Bioassay is sensitive but specific only for various unconjugated pteridines and also time-consuming. Radioimmunoassay is sensitive, specific and convenient for screening, but requires expensive reagents and involves the handling of radioactive materials.

We have developed a nonisotopic immunoassay for biopterin and neopterin, using a new polarization fluoroimmunoassay technique, and have applied this method to the determination of these pterins in human urine (32).

Preparation of Fluorescein-labelled Biopterin and Neopterin (Fig. 1)

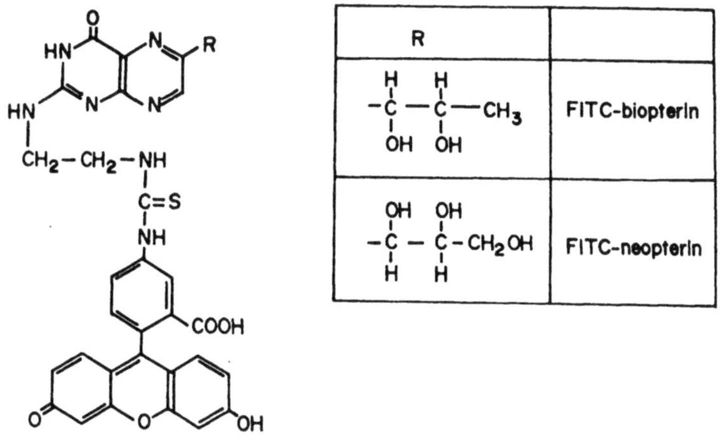

Fig. 1. Structure of fluorescein-labelled L-erythro-biopterin and D-erythro-neopterin.

A solution of 4-hydroxy-6-(L-erythro-1,2-dihydroxypropyl)-2-methylthiopteridine (31) (1.0 g) and 1,3-diaminopropane (1.0 g) in 50% aqueous 2-methoxyethanol (20 ml) was heated under reflux for 2 hr. The solution was adjusted to pH 2-3 with formic acid and fractionated on a Florisil column (83.5 x 40 cm), using 3% ammonia as solvent. The eluate was evaporated to dryness and the residue was chromatographed on a Dowex 50W X8 column (2 x 8 cm), eluted gradiently by 0-5% ammonia (31). Evaporation of the eluate gave a yellow viscous oil (0.9 g) which was chromatographically pure but did not crystallize. Part of the oil (0.1 g) was treated with acetic anhydride and pyridine in a conventional way to give ivory needles (80% yield) of 2-(3-acetyl-aminopropyl)-amino-4-hydroxy-6-(L-erythro-1, 2-dihydroxy-propyl)-pteridine, m.p. 232°C-234°C, decomp. (recrystallized from ethanol) (Found: C, 50.52; H, 6.10; N, 24.85%. Calculated for $C_{14}H_{20}N_6O_4$: C, 50.00; H, 5.95; N, 25.00%); λmax (logϵ) in methanol: 238 (shoulder, 4.01), 280 (4.30), 352 (3.74)nm.

A solution of the above oil (0.5 g), fluorescein isothiocyanate (0.5 g), and potassium carbonate (0.2 g) in 20% aqueous methanol (30 ml) was kept at room temperature for 10 hr.

The solution was adjusted to pH 3 with formic acid and chromatographed on a silica gel column (Walo Gel C-100, 3 x 40 cm) using a mixture of acetic acid, water and 1-butanol (1:1:4) as solvent. The eluate was evaporated to dryness and the residue was crystallized from methanol to give orange coloured needles (0.21 g) of the fluorescein-labelled biopterin, which did not melt below 300°C (Found: C, 57.60; H, 4.55; N, 14.65%. Calculated for $C_{33}H_{29}N_7O_8S$: C, 57.97; H, 4.28; N, 14.34%).

Aminolysis of 4-hydroxy-6-(D-erythro-1,2,3,-trihydroxypropyl)-2-methylthiopteridine with 1,3-diaminopropane gave the 2-(3-aminopropyl)-aminopteridine (75% yield), which was similarly characterized as an acetamide. The amide melted at 205°C-207°C, decomp. (recrystallized from methanol) (Found: C, 47.52; H, 5.73; N, 23.81%. Calculated for $C_{14}H_{20}N_6O_5$:C, 47.73; H, 5.86; N, 23.86%) and exhibited λmax (logϵ) in methanol at 237 (shoulder, 4.01), 280 (4.30) and 352 (3.74) nm. Condensation of the amine with fluorescein isothiocyanate and isolation of the product in an analogous way, as above, gave the fluorescein-labelled neopterin as orange coloured needles (30% yield), m.p. not below 300°C (Found; C, 56.86; H, 4.46; N, 14.34%. Calculated for $C_{33}H_{29}N_7O_9S$: C, 56.65; H, 4.18; N, 14.01%).

Preparation of Antiserum and Standard

Antisera of biopterin and neopterin were raised in rabbits and characterized as previously described (25-27, 30). Reference standards of biopterin and neopterin were synthesized and purified as previously reported (25-27, 30).

Specimens

Since human urine contains dihydro- and tetrahydro-derivatives of biopterin and neopterin (21), the reduced derivatives of both biopterin and neopterin were oxidized by iodine in 0.2M HCl to the parent pteridines before the assay. To 500 µl of urine, a 1/10 volume of 2M HCl and 1/10 volume of 2% I_2-4% KI solution were added in this order and the solution was left for 1 hr at room temperature in the dark. Oxidation was terminated by adding a 1/10 volume of a 2% ascorbate solution and the mixture was lyophilized. The residue was dissolved in 500 µl of 0.02M phosphate buffer (pH 7.5), and used for the fluoroimmunoassay.

Fluorometers

Two fluorescence depolarization analyzers, JASCO SFP-3 (Japan Spectroscopic Co., Hachioji, Tokyo 192, Japan) and FS-501 (Union Giken Co., Hirakata, Osaka 573-01, Japan) were used.

Assay Procedure

During the entire assay procedure, 0.02M phosphate buffer (pH 7.5) was used. For routine assay, 100 µl of the sample, or 100 µl of the buffer containing standard biopterin or neopterin (from 0.1 to 10 nmol), and 100 µl of a FITC-biopterin or FITC-neopterin solution (3 mg/l; the final pterin content: 100 ng/ml) were mixed with 2.7 ml of the buffer, 100 μl of 100-fold diluted antiserum was then added to give a dilution of 1:3000 in a final volume of 3.0 ml. After incubation at 25°C for at least 2 min, the fluorescence polarization was calculated by using the equation: percent polarization = $(P - P_{blank}/P_0 - P_{blank}) \times 100$, where P = polarization of each sample, P_{blank} = polarization of a FITC-pterin alone, and P_0 = polarization in the absence of a pterin. The percent polarization was plotted against the logarithm of the concentration of the standard pterins, and the biopterin or neopterin levels in urine samples were determined from this standard curve.

Results

Figure 2 shows the antibody dilution curve of biopterin and neopterin, where the concentrations of FITC-biopterin and FITC-neopterin were kept constant at 100 ng/ml. The antisera of

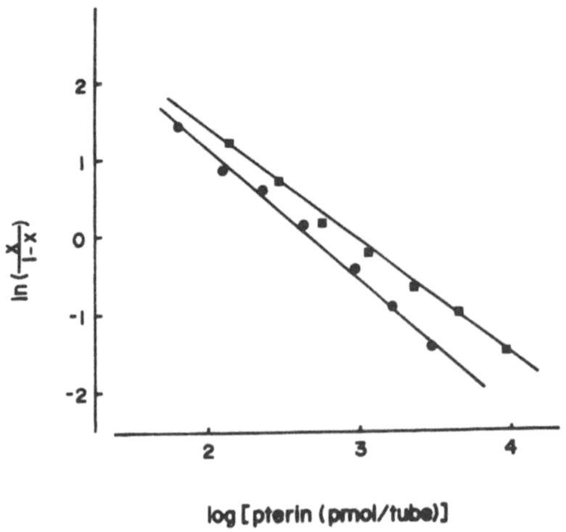

Fig.3. Standard curves showing decreasing percent polarization with increasing levels of added biopterin (■—■) and neopterin (●—●).

biopterin or neopterin showed about 30% binding of FITC-biopterin and FITC-neopterin, respectively, at a dilution of 1:3000 in final concentrations. In Fig. 3, typical standard curves for the assay of biopterin and neopterin are indicated. In these

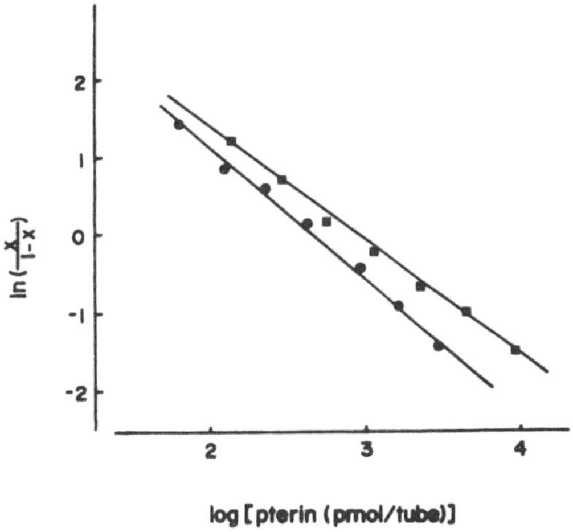

Fig.3. Standard curves showing decreasing percent polarization with increasing levels of added biopterin (■—■) and neopterin (●—●).

conditions, sensitivities of biopterin and neopterin are 0.05 nmol/tube and 0.03 nmol/tube, respectively. The two FITC-pterins gave similar curves at the same concentrations of biopterin and neopterin. It was found that the standard curve shifted to lower concentrations when the concentration of FITC-pterins was lowered, as illustrated in Fig. 4 in the case of neopterin.

The concentration to the fluorescence polarization signal made by the intrinsic fluorescence of urine samples or standards was found to be negligible, therefore, the blank value (P_{blank}) was given by fluorescence polarization of FITC-pterins alone.

When the minimum incubation time was assessed by adding antisera to FITC-biopterin and FITC-neopterin in the presence

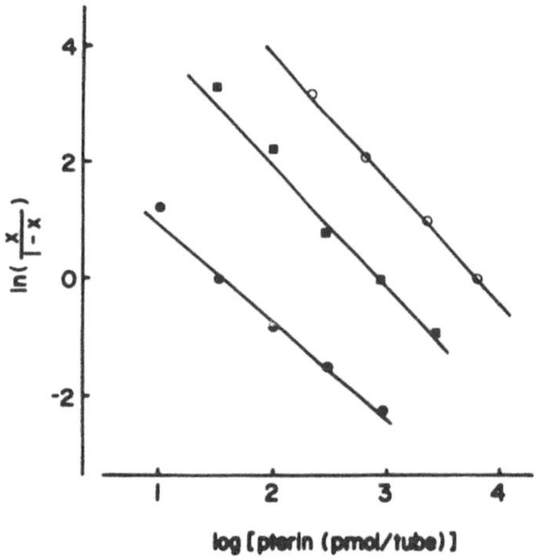

Fig.4. Effects of FITC-neopterin concentration on standard curve of neopterin. O—O, antiserum 1:2.4 x 10^4, FITC-neopterin 25 ng/tube; ■—■, antiserum 1:3.2 x 10^4, FITC-neopterin 20 ng/tube; ●—●, anti-serum 1:3.2 x 10^5, FITC-neopterin 2.5 ng/tube.

of various amounts of biopterin and neopterin, equilibrium in the increase in fluorescence polarization was reached within 2 min at 25°C, and the polarization reading became stable thereafter.

Linearity of assay for the pterins in human urine was obtained from 50 µl up to 500 µl urine, and recoveries of authentic biopterin and neopterin added to urine samples were 96% and 95% according to this method. The inter-assay variation (C.V.) with replicates of the same urine sample was within 6.0% (n=12) and precision was satisfactory.

Table I shows the biopterin and neopterin contents of urine samples measured by polarization fluoroimmunoassay. Normal ranges of pterins in human urine samples are 0.5-5 nmol/ml urine in neopterin and 0.5-3 nmol/ml urine in biopterin. The correlation between fluoroimmunoassay and radioimmunoassay (23, 24) was satisfactory. The correlation coefficients (r) between the two assays of neopterin and biopterin were 0.93 ($p<0.01$, n=23) and 0.96 ($p<0.01$, n=20), respectively, as illustrated in Fig. 5A and B.

TABLE I Pterin content in human urine samples

Pterin content (nmol/ml)	
neoptein	biopterin
5.73	3.49
0.41	2.20
1.40	1.37
0.71	1.67
2.13	2.00
1.73	1.96
1.48	2.20
3.96	2.56
0.57	0.97

Conclusion

The polarization fluoroimmunoassay of urinary biopterin and neopterin has several practical advantages over established assay methods. The assay is fast, it takes only 2 min to read out the fluorescence polarization after mixing the sample, FITC-pterins and antiserum. It is economical, and the specificity and accuracy are satisfactory. Furthermore, it does not require a separation step. One drawback is its relatively low sensitivity, approximately 30 pmol/tube, as compared to the sensitivity of radioimmunoassay (0.1 pmol/tube) (26, 30) or bioassay (0.05 pmol/ml) (24), but it is sensitive enough to measure biopterin and neopterin in human urine and, therefore, it is useful for the screening of the two pterins in urine samples from patients. The method is also ideally suited to total automation by the continuous-flow system.

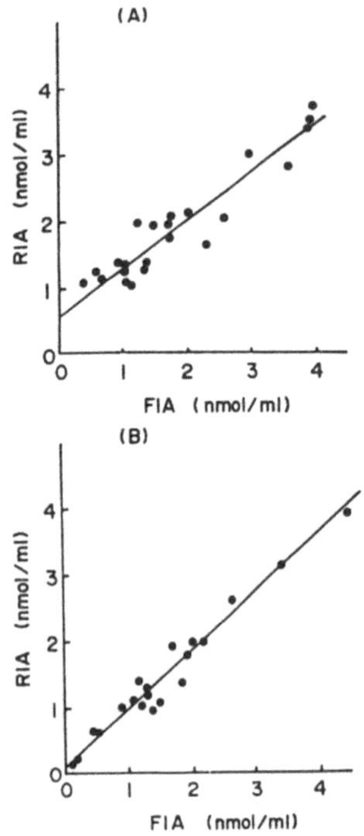

Fig.5 (A) Correlation between neopterin levels in human urine samples as determined by polarization fluoroimmunoassay (FIA) and those by radioimmunoassay (RIA). $r=0.93$, $p<0.01$, $n=23$. ——, line of best fit. (B) Correlation between biopterin levels in human urine samples as determined by FIA and those by RIA. $r=0.96$, $p<0.01$, $n=20$. ——, line of best fit.

References

1. O'Donnell, C.M., Suffin, S.C.: Fluorescence immunoassay. Anal. Chem. 51, 33-40 (1979).
2. Curry, R.E., Heizman, H., Riege, D.H., Sweet, R.V., Simonsen, M.G.: A systems approach to fluorescent immunoassay: General principles and representative applications. Clin. Chem. 25, 1591-1595 (1979).
3. Watson, R.A.A., Landon, J., Show, E.J., Smith, D.S.: Polarisation fluoroimmunoassay of gentamicin. Clin. chim. Acta 73, 51-55 (1976).

4. Kaufman, S.: Studies on the mechanism of the enzymatic conversion of phenylalanine to tyrosine. J. biol. Chem. 234, 2677-2682 (1959).

5. Kaufman, S.: The structure of the phenylalanine-hydroxylation cofactor. Proc. natn. Acad. Sci. U.S.A. 50, 1085-1093 (1963).

6. Nagatsu, T., Levitt, M., Udenfriend, S.: Tyrosine hydroxylase. The initial step in norepinephrine biosynthesis. J. biol. Chem. 239, 2910-2917 (1964).

7. Brenneman, A.R., Kaufman, S.: The role of tetrahydropteridine in the enzymatic conversion of tyrosine to 3,4-dihydroxyphenylalanine. Biochem. biophys. Res. Commun. 17, 177-183 (1964).

8. Lovenberg, W., Jequire, E., Sjoerdsma, A.: Tryptophan hydroxylation: Measurement in pineal gland, brain stem and carcinoid tumor. Science 155, 217-219 (1967).

9. Ichiyama, A., Nakamura, S., Nishizuka, Y., Hayaishi, O.: Enzymic studies on the biosynthesis of serotonin in mammalian brain. J. biol. Chem. 245, 1699-1709 (1970).

10. Friedman, P.A., Kappelman, A.H., Kaufman, S.: Partial purification and characterization of tryptophan hydroxylase from rabbit hind brain. J. biol. Chem. 247, 4165-4173 (1972).

11. Kaufman, S., Berlow, S., Summer, G.K., Milstien, S., Schliman, J.D., Orloff, S., Spielberg, S., Pueschel, S.: Hyperphenyl alaninemia due to a deficiency of biopterin. A variant form of phenylketonuria. New Engl. J. Med. 299, 673-679 (1978).

12. Curtius, H.Ch., Niederwisser, A., Viscontini, M., Otten, A., Schaub, J., Scheibenreiter, S., Schmidt, H.: Atypical phenylketonuria due to tetrahydropterin deficiency. Diagnosis and treatment with tetrahydrobiopterin, dihydrobiopterin and sepiapterin. Clin. chim. Acta 93, 251-262 (1979).

13. Nixon, J.C., Lee, C.L., Milstien, S., Kaufman, S., Bartimore, K.: Neopterin and biopterin levels in patients with atypical phenylketourea. J. Neurochem. 35, 898-904 (1980).

14. Lovenberg, W., Levine, R.A., Robinson, D.S., Ebert, M., Williams, A.C., Calne, D.B.: Hydroxylase cofactor activity in cerebrospinal fluid of normal subjects and patients with Parkinson's disease. Science 204, 624-626 (1979).

15. Nagatsu, T., Yamaguchi, T., Kato, T., Sugimoto, T., Matsuura, S., Akino, M., Nagatsu, I., Iizuka, R., Narabayashi, H.: Biopterin in human brain and urine from controls and parkinsonian patients: application of a new radioimmunoassay. Clin. chim. Acta 109, 305-311 (1981).

16. Plowman, J., Cone, J.E., Guroff, G.: Identification of D-erythro-dihydroneopterin triphosphate, the first product of pterine biosynthesis in Comamonus Sp. (ATCC 11299a). J. biol. Chem. 249, 5559-5564 (1974).

17. Eto, I., Fukushima, K., Shiota, T.: Enzymatic synthesis of biopterin from D-erythro-dihydroneopterin triphosphate by extracts of kidneys from Syrian hamsters. J. biol. Chem. 251, 6505-6512 (1976).

18. Fukushima, K., Richter, W.E., Shiota, T.: Partial purification of 6-(D-erythro-1',2',3'-trihydroxypropyl)-7,8-dihydropterin triphosphate synthetase from chicken liver. J. biol. Chem. 252, 5750-5755 (1977).

19. Kapatos, G., Katoh, S., Kaufman, S.: Biosynthesis of biopterin by rat brain. J. Neurochem. 39, 1152-1162 (1982).

20. Rocos, H., Rocos, K., Kirstaedter, H.-J.: Altered urinary excretion of pteridines in neoplastic disease. Determination of biopterin, neopterin, xanthopterin and pterin. Clin. chim. Acta 105, 275-286 (1980).

21. Stera, B., Halpern, R.M., Halpern, B.C., Smith, R.A.: Urinary excretion levels of unconjugated pterins in cancer patients and normal individuals. Clin. chim. Acta 113, 231-242 (1981).

22. Fukushima, T., Nixon, J.C.: Analysis of reduced forms of biopterin in biological tissues and fluids. Anal. Biochem. 102, 176-188 (1980).

23. Lunte, C.E., Kissinger, P.T.: Determination of pterins in biological samples by liquid chromatography/electro-chemistry with a dual-electrode detector. Anal. Chem. 55, 1458-1462 (1983).

24. Dewey, V.C., Kidder, G.W.: Assay of unconjugated pterins. In: "Method in Enzymology", Eds. McCormick, D.B., Wright, L.D., Academic Press, New York, 18B, pp. 618-629 (1971).

25. Nagatsu, T., Yamaguchi, T., Kato, T., Sugimoto, T., Matsuura, S., Kobayashi, K., Akino, M., Tsushima, S., Nakazawa, N., Ogawa, H.: Radioimmunoassay for biopterin. Proc. Japan Acad. 55, ser. B., 317-322 (1979).

26. Nagatsu, T., Yamaguchi, T., Kato, T., Sugimoto, T., Matsuura, S., Akino, M., Tsushima, S., Nakazawa, N., Ogawa, H.: Radioimmunoassay for biopterin in body fluids and tissues. Anal. Biochem. 110, 182-189 (1981).

27. Nagatsu, T., Sawada, M., Yamaguchi, T., Sugimoto, T., Matsuura, S., Akino, M., Nakazawa, N., Ogawa, H.: Radioimmunoassay for neopterin in body fluid and tissues. In: "Chemistry and Biology of Pteridines", Ed. Blair, J.A., Walter de Gruyter, Berlin, New York, pp. 821-825 (1983).

28. Rocos, H., Rocos, K.: A radioimmunoassay for determination of D-erythro-neopterin. In: "Chemistry and Biology of Pteridines", Ed. Blair, J.A., Walter de Gruyter, Berlin, New York, pp. 815-819 (1983).

29. Sugimoto, T., Matsuura, S., Yamaguchi, T., Sawada, M., Nagatsu, T.: Determination of biopterin and neopterin in

fluid and tissues by radioimmunoassay. In: "Biochemical and Clinical Aspects of Pteridines" vol. 2, Eds. Curtius, C.H., Pfleiderer, W., Wachter, H., Walter de Gruyter, Berlin, New York, pp. 53-63 (1983).

30. Nagatsu, T., Sawada, M., Yamaguchi, T., Sugimoto, T., Matsuura, S., Akino, M., Nakazawa, N., Ogawa, H.: Radioimmunoassay for neopterin in body fluids and tissues. Anal. Biochem. (in press) (1984).

31. Sugimoto, T., Matsuura, S., Nagatsu, T.: Studies on biological active pteridines IV. Bull. Chem. Soc. Jpn. 53, 2344-2347 (1980).

32. Sawada, M., Yamaguchi, T., Sugimoto, T., Matsuura, S., Nagatsu, T.: Polarization fluoroimmunoassay of biopterin and neopterin in human urine. Clin. chim. Acta 138, 275-282 (1984).

LUMINESCENCE IMMUNOASSAYS IN THEORY AND PRACTICE - THE STATE OF THE ART

W. G. Wood
Klinische Laboratorien, Klinik für Innere Medizin, Medizinische Hochschule Lübeck, Ratzeburger Allee 160, 2400 Lübeck 1, Federal Republic of Germany

Introduction

The aim of this review is to present the current state of the luminescence immunoassay (LIA) as a singular alternative to the radioimmunoassay (RIA). The accent is placed more upon the practical applications and methodological details rather than on the theoretical considerations, about which many publications have appeared over the past decade, including several excellent review articles (1-9).

Many of the examples used in this article have been taken from everyday routine LIAs which are performed in the author's laboratory. In addition, assays will be presented from the major internationally known laboratories working in this field. The number of laboratories actively working on luminescence immunoassays is still small but there is an exchange of data and reagents, the latter being important as the number of commercially available products is still inadequately small.

Luminescence and the closely related fluorescence and phosphorescence can be defined as follows.

a) Luminescence

Luminescence is the result of highly exergonic reactions in which energy (between 40 and 70 kcal/mol for visible light) is released as light quanta. In most cases, nascent oxygen plays

Immunoassay Technology, Volume 1

an important role with dioxetanones being formed as intermediates (10, 11). Luminescence is the only case where chemical energy is converted into light quanta, the efficiency of this process being from under 1% for many chemiluminescent reactions to above 90% for certain bioluminescent systems. Further physical details as to how luminescence "works" are already available (5).

b) Fluorescence

Fluorescence is produced when a substance is irradiated with light of a given wavelength, and as a response, light of a longer wavelength (lower energy) is emitted. In contrast to luminescence, no chemical reaction takes place and a primary light source is needed. The relationship between the excitation and emission wavelengths is governed by Stokes' law. Fluorescence occurs when a molecule returns from the first singlet (S_1) excitation level directly to the ground state (S_0).

c) Phosphorescence

Phosphorescence is a special case of fluorescence in which a molecule returns from the first triplet (T_1) excitation level directly to the ground state (S_0). In contrast to fluorescence, phosphorescence is usually "long-lived", i.e. the release of energy from the triplet state is much slower than from the singlet state. There is a "mixed" state of affairs, called delayed fluorescence or alpha phosphorescence in which the molecule passes from the first triplet state to the ground state via the first singlet state. For more details on fluorescence and phosphorescence, the reader is referred to the work of Barenboim, Domanskii and Turoverov (9).

A Short History of Luminescence

The first reference to luminescence is to be found in a Chinese manuscript written some 3,500 years ago which described the light of fireflies or glow-worms. The phenomenon of luminescence has interested many people including Aristotle, Gaius Plinius Secundus, Francis Bacon and Leonardo da Vinci, although Robert Boyle was the first to lay down the basic tenets of bioluminescence, which are still valid today, over 300 years after their presentation to the Royal Society in London (12, 13). The terms luciferin and luciferase were coined by Dubois (14, 15), but chemiluminescence is of more recent origin.

The field of luminescence immunoassays is confined to the past decade and the only way in which the LIA differs from the RIA is in the choice of signal used. The next two headings deal with the choice of label - either bioluminescent or chemiluminescent - and the advantages and disadvantages of both groups of substances.

Bioluminescence

As mentioned above, the principles of bioluminescence have already been described (12, 13). The main discoveries in bioluminescence since Boyle's time have been made by Dubois with his discovery of the enzyme-substrate luciferase-luciferin system (14, 15) and by McElroy who made a detailed study of luciferin-luciferase interactions (16, 17), especially those involving ATP, the basic features of which are shown in Fig. 1. A second NAD(P)H-dependent luciferase-luciferin system is found in certain luminescent bacteria, where long-chain aliphatic aldehydes often play the part of a luciferin, the role of the luciferase being taken by a dehydrogenase. Figure 2 shows the reaction scheme of the NAD(P)H system in a simplified form.

The original role of bioluminescence may simply have been a detoxification mechanism to eliminate oxygen radicals. This is supported by the fact that bioluminescence is confined to lower organisms and is absent in reptiles, birds, marsupials and mammals as well as in the more evolved members of the plant

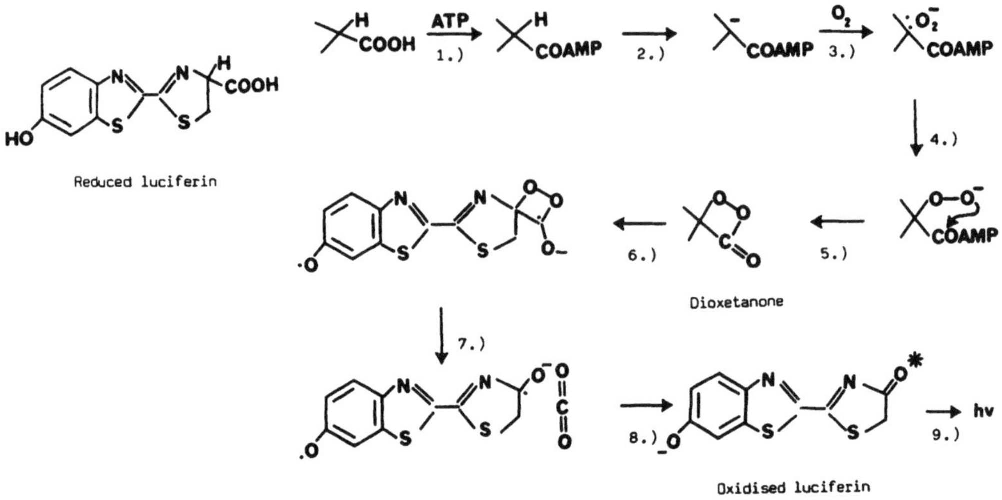

Fig. 1. The reaction sequence of firefly-luciferin with ATP and luciferase to give oxidised luciferin and light (hv).

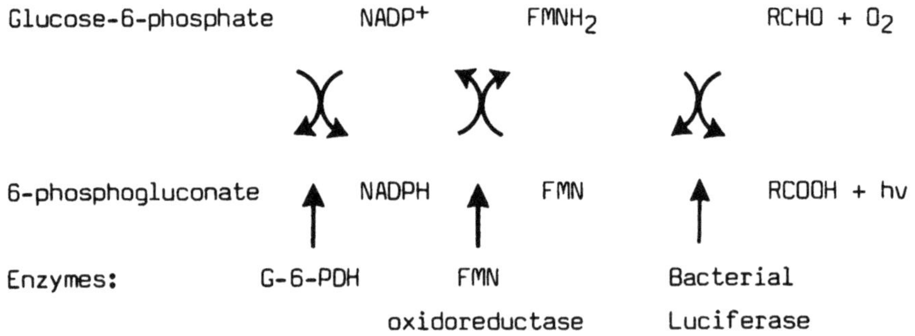

Fig. 2. The bacterial luciferin-luciferase reaction coupled to a dehydrogenase system. G-6-PDH=glucose-6-phosphate dehydrogenase; FMN/$FMNH_2$=flavin mononucleotide in oxidised and reduced forms; $NADP^+$/NADPH=nicotine adenine dinucleotide phosphate - oxidised and reduced; RCHO=aliphatic aldehyde, e.g. n-decanal; RCOOH=corresponding carboxylic acid, e.g. n-decanoic acid; hv=light.

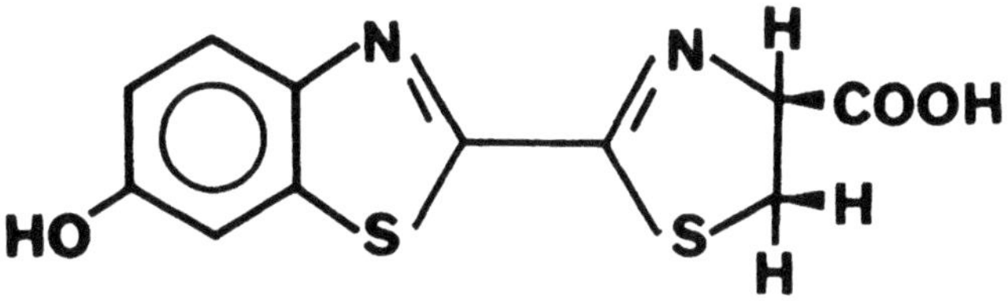

1

$CH_3(CH_2)_nCHO \quad n = 8-12$

2

Fig. 3. Structure of firefly luciferin (1) and bacterial luciferin (2).

kingdom (Fig. 3). The colour of the light emitted is dependent upon the luciferase (enzyme) rather than the luciferin (substrate) which shows that energy-transfer systems exist in nature to modify the primary luminescence of the luciferin, which is almost always in the blue region of the spectrum.

Adaption of bioluminescent systems for in vitro use presents problems as the enzyme systems concerned are often sensitive to substances such as merthiolate (organomercury) and azide which are often used as anti-microbial agents. This makes storage in liquid form difficult except under sterile conditions, or in the presence of antimycotica and bacteriostatica such as aminoglycosides, penicillin or similar substances. This stability/storage problem frequently outweighs the high efficiency of the light-generating systems and is the reason why the assays in the laboratory of the author, although at first using bioluminescence (18, 19), now use a chemiluminescent system, which, despite the low quantum efficiency of the luminescent reaction (under 5%), is stable enough to allow storage of reagents under normal laboratory conditions (i.e. at $4^{o}C$ or $-20^{o}C$) over extended periods of time (as long as two years). The advantages of bioluminescent systems in in vitro systems are: 1) High quantum efficiency - up to 90% of the theoretical light emission, 2) kinetic measurements allowed - remeasurements possible, 3) amplification of signal possible - by use of kinases and dehydrogenases as labels. Disadvantages are: 1) Systems often susceptible to poisons and ionic effects - e.g. azide, merthiolate, thiocyanate, divalent cations, 2) storage of reagents in liquid form difficult, 3) cost of reagents high, and 4) synthesis of labels often complicated.

Chemiluminescence

An early report on chemiluminescence indicated that light was emitted during oxidation of lophine (2,4,5 triphenyl imidazole) (20) but later publications (21, 22) made observations on luminol and lucigenin which opened the way for modern chemiluminescence.

The majority of luminescence immunoassays and phagocytosis set-ups use luminol or a luminol-derivative as signal (23, 24). Other tracer-molecules include acridine derivatives - mainly acridinium esters (25, 26) and fluorescein derivatives (27). Other compounds are aryl oxalate esters, which not only have a quantum efficiency of above 0.3 (30%), but have a light output lasting for several hours, hence their use in "cold-light" sticks (28). The basic structures of these compounds are shown in Fig. 4.

Although a luciferase is not needed for chemiluminescence, the luminol-based systems use a peroxidase or pseudoperoxidase (haem-group containing moiety) to generate the active oxygen species. Examples of pseudoperoxidases include cytochrome-c and microperoxidase, both of which are available commercially. The term pseudoperoxidase is apt, because the molecule does not

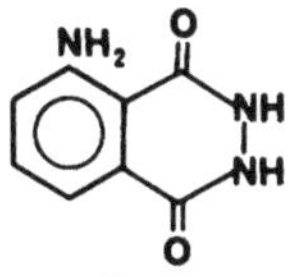

a

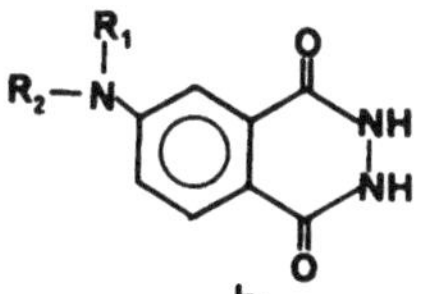

b

1

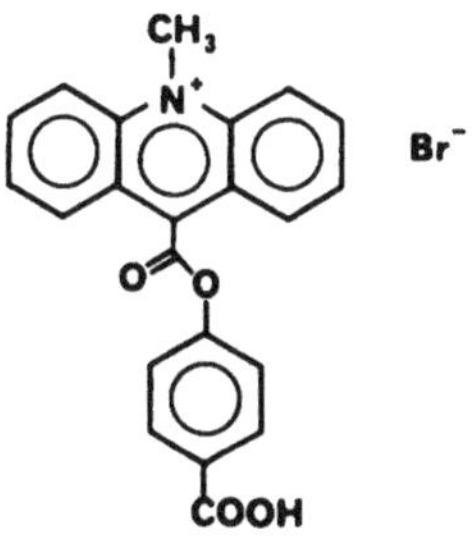

2

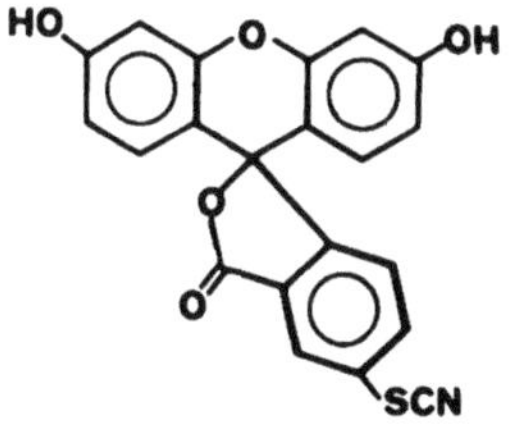

3

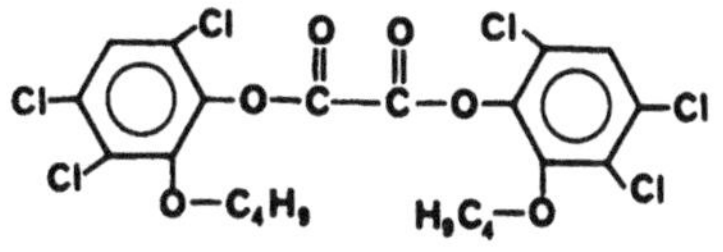

4

Fig. 4. Structure of four chemiluminescent label groups: 1a. Luminol (5-amino-2,3-dihydro-1,4-phthalazinedione). 1b. Isoluminol group. R_1=H, R_2=H - isoluminol (6-amino-2,3-dihydro-1,4-phthalazinedione); R_1=ethyl, R_2=4-aminobutyl - ABEI; R_1=ethyl, R_2=5-aminopentyl - APEI; R_1=ethyl, R_2=6-aminohexyl - AHEI; R_1=ethyl, R_2=N-(4-aminobutyl)-hemisuccinamide - ABEI-H. 2. Acridinium ester group, e.g. 9-methylacridine-10 carboxy-(4-hydroxybenzoate). 3. Fluorescein group, e.g. fluorescein isothiocyanate. 4. Aryl oxalate ester group, e.g. bis(2,4,5-trichloro-6-butoxy)-oxalate.

act as a true enzyme as it only reacts once and not in a cyclic fashion during light generation. This point is important to note in the oxidation systems using haem-containing compounds, as claims of amplification of signal using antigen or antibody labelled with peroxidase must be discredited, when the latter are not used exclusively to remove peroxide from the system.

The efficiency of chemiluminescence is low, ranging from under 1% for lucigenin (29) to above 50% for certain acridinium esters (30). In comparison, the firefly-ATP bioluminescent system has a quantum efficiency of around 90%. The light-output of chemiluminescent systems is pH-dependent both in terms of yield and kinetics. The advantages of chemiluminescence when applied to _in vitro_ systems are: 1) Reagents stable, even in the presence of azide, for long periods of time, 2) reagent costs relatively low, 3) molecules small enough not to influence antigen or antibody structure in most cases, 4) full control of the light emission reaction - in contrast to radioactivity where one is dependent upon the decay constant of the nuclide in question. Disadvantages are: 1) Low quantum efficiency - often below 5% of the theoretical yield, 2) usually no chance of a remeasurement of a sample, 3) suffers from the same disadvantages as β-scintillation counting - i.e. quenching.

Labels

There are two classes of label, those for use in bioluminescent systems, i.e. a luciferin, luciferase or co-factor, or an enzyme which generates ATP (a kinase, 18), or NAD(P)H (a dehydrogenase, 31), and labels for use in chemiluminescent systems, for example, luminol-, acridinium- or oxalate-ester derivatives (24, 25, 32). It is possible to use an indirect-labelling system where a two-component system such as avidin-biotin is a good example (33), this being a method to keep the label and serum components apart during the early incubation stages (see later under assay principles).

Examples of labels used in bioluminescence include pyruvate kinase (18) for ATP-dependent systems and glucose-6-phosphate dehydrogenase (G-6-PDH) for NAD(P)H-dependent luciferase-luciferin systems.

In chemiluminescent systems two types of label have become accepted, the isoluminol derivatives, for example ABEI (6-(4-aminobutyl-N-ethyl)-isoluminol), which are stable but have a lower quantum efficiency, and the acridinium esters (25, 34), which have a higher quantum efficiency but at the same time show a stability which is very pH dependent (35). Because of their structures, chemiluminescent labels such as those mentioned above can be stored in azide-containing buffers and can be used in assay-buffer systems already optimised for RIA, which often contain bacteriostatica and detergents of the Tween, Brij or Triton families. Systems using labelled co-factors such as NAD (33), ATP (36) or peroxidase (37) are described, but have not managed to displace the conventional labelling of antibody or antigen with the luminogen.

Oxidation Systems

As stated above, the light production is dependent upon an oxidation step, therefore the oxidation system used is of utmost importance in optimising a luminescent immunoassay. Each luminogen has its own optimal oxidation system, i.e. there is no universal system *per se*.

The phthalhydrazide group (luminol/isoluminol and derivatives) requires a haem-group in the form of a (pseudo) peroxidase, a supplier of nascent oxygen (peroxide, perborate or persulphate) and usually (but not always) an alkaline reaction medium (sodium triphosphate, sodium hydroxide, sodium tetraborate). In general, the higher the pH (in direction pH 14), the higher the light output and the slower the kinetics when dealing with this group of luminogens (see Fig. 5).

Acridinium esters only require alkaline peroxide for the

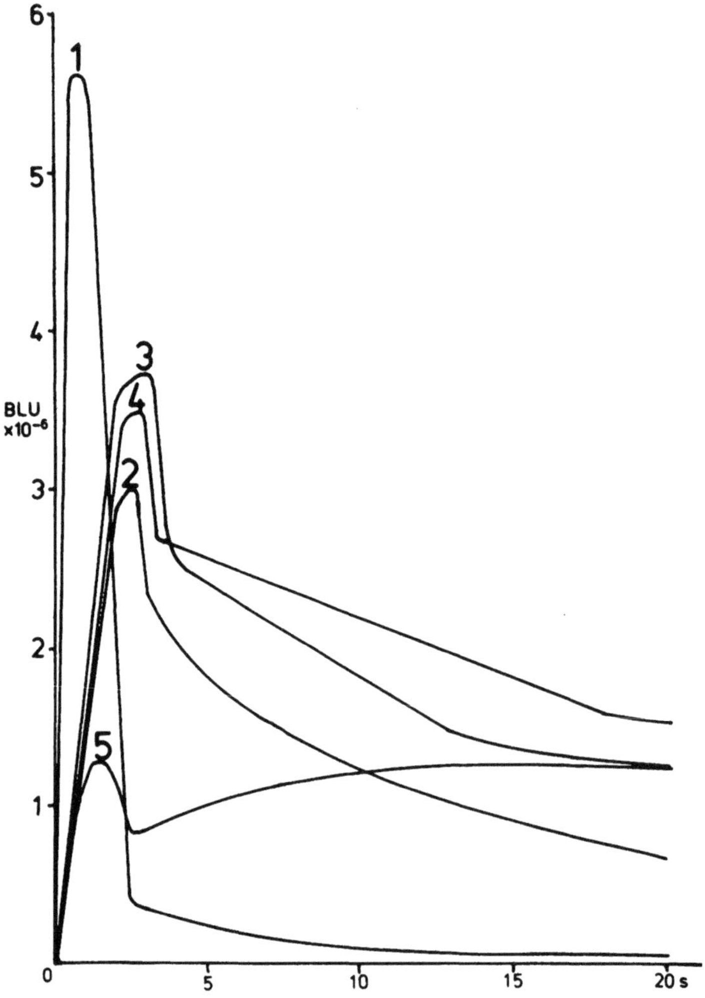

Fig. 5. Light emission curves at different pH-values. The oxidation system is as in reference 42, but where the 1 mol/l sodium hydroxide is replaced with the solution under test. Curve 1 with 0.1 mol/l phosphate, pH 8.6, curves 2-5 with 0.25, 0.5, 1 and 5 mol/l sodium hydroxide respectively. The luminogen used was ABEI-I labelled rabbit-anti transferrin. The areas under the 20 s integral for the label and reagent blank were as laid out below in thousand Berthold light units (BLU). These units are arbitrary and are related to the number of photons emitted.

Solution under test	Specific signal	Reagent blank
0.1 mol/l phosphate	9200	216
0.25 mol/l NaOH	19500	715
0.50 mol/l NaOH	22000	1200
1.00 mol/l NaOH	23500	58
5.00 mol/l NaOH	14900	19

This figure demonstrates the need to optimise each system according to the kinetics required.

light production (38) which makes their use attractive. One must not forget that the lower the activation energy needed the less stable the compound, and it is a known fact that acridinium esters are not as stable as phthalhydrazides so that one must know how to use and store acridinium-ester labelled compounds.

Fluorescein derivatives require only hypochlorite solution to generate light, and although this appears to be an advantage, it has often been found that the stainless steel injection needles and pump components coming into contact with hypochlorite are rapidly attacked, leading to costly repairs.

Aryl oxalate esters have not been tried fully as labels in luminescence immunoassays and many of the oxidation systems described are in organic solutions, which prevent the use of polystyrene or polycarbonate assay tubes. The high quantum efficiency of these oxalate esters (up to 0.35), is probably enough of a stimulus to develop acceptable oxidation systems and can be developed for immunoassay purposes.

Ways of increasing the light output have not only been published but also patented, especially for the enhancement of light output in a luminol-system using horeseradish peroxidase in combination with d-luciferin as amplifier (41). It has been noticed by the author that, in a similar system, but where microperoxidase (MP 11 - Sigma - Catalogue No. M-6756) is used instead of the less efficient (but less expensive) horseradish peroxidase, premixing the sodium hydroxide and hydrogen peroxide leads to an increase in light output over the first 15-30 min after mixing, the increased light output being stable over the subsequent 24 hr-36 hr (42). The latter effect is due to production of OOH^--ions (29), the former due to energy-transfer processes not yet fully understood. Table 1 shows a few published oxidation systems for chemiluminescent immunoassays together with references.

Table 1. Some oxidation systems for chemiluminescence immunoassays.

Reference	Label	Oxidation system	Integration
Shapiro et al (27)	FITC	Na or Ca hypochlorite	0 - 30 sec
Hersch et al (23)	Luminol	Hemin/peroxide/pH 11	Peak height
Kim et al (49)	ABEI	MP-11/peroxide/pH 14	0 - 10 sec
Kohen et al (48)	APEI	MP-11/peroxide/pH 8.6	2 - 12 sec
Wood et al (42)	ABEI-H	MP-11/peroxide/pH 13	0 - 20 sec
Whitehead et al (41)	HRP	Luciferin/luminol/pH 8	0 - 30 sec
Tsuji et al (32)	GOD	TCPO/ANS	15 - 21 sec
Weeks et al (38)	AcEs	NaOH/peroxide	0 - 2 sec

Fluorescein isothiocyanate (FITC), horseradish peroxidase (HRP), glucose oxidase (GOD), acridinium ester (AcEs), microperoxidase (MP-11), 8-anilino-1-naphthalene sulphonic acid (ANS), bis-(2, 4,6-trichlorophenyl) oxalate (TCPO), 6-(4-aminobutyl-N-ethyl) isoluminol (ABEI), 6-(5-aminopentyl-N-ethyl) isoluminol (APEI), 6-(N-(4-aminobutyl-N-ethyl) isoluminol) hemisuccinamide (ABEI-H).

Assay Systems

The choice of assay system is similar to that for radioimmunoassay, using either liquid or solid-phase assays. Although the use of solid-phases is preferred for reasons described below, there are liquid-phase assays which perform very well, provided that an extraction step precedes the actual assay. Liquid-phase assays have successfully been developed for steroids and their metabolites in urine, serum and plasma, and examples of such systems are given later.

The terminology used here is derived from RIA and the abbreviated terms are mentioned in parentheses, as follows.

Chemiluminescent immunoassay using a labelled antigen (CELIA); immunoluminometric assay with two substance-specific antibodies, one on the solid-phase and one labelled with the luminogen (ILMA); solid-phase antigen luminescence techniques - where the antigen or in the case of haptens an antigen-protein conjugate, is coupled to the solid phase, the label is a labelled species-specific (second) antibody (SPALT); immunoluminometric labelled second-antibody assay (ILSA), which is similar to the (IRMA) with a "second" antibody tracer; bioluminescent labelled immunoassays where an enzyme has been coupled to the antigen (EELIA): enzyme enhanced luminescence immunoassay, and where labelled co-factors have been used (LUCIA): luminescent co-factor immunoassay. The EELIA is an extension of the enzyme immunoassay. Figures 6a-6d show the assay principles of CELIA, ILMA, ILSA and SPALT techniques. The EELIA and LUCIA methods are analogous to the CELIA, only the mode of labelling the antigen being different.

The advantages and disadvantages of each system are listed as follows. The advantage of CELIA is that solid or liquid-phase antibodies can be used in an analogous way to radioimmunoassay. Disadvantages are that 1) label and sample come into contact with each other giving rise to possible degradation of label through action of sample components, 2) quench effects present in coloured (yellow)samples, 3) extraction step necessary to reduce 1) and 2) above, 4) synthesis of antigen-luminogen often difficult and must be carried out for each antigen. Advantages of ILMA are 1) label and sample incubations can be kept separate thus reducing potential interference, 2) quench effects through 1) above not present, 3) "antibody-excess" conditions reduce assay time and can increase the dynamic range of an assay, 4) labelling of antibodies without loss in immunoreactivity simple. Disadvantages are that 1) antigen must have two independent antibody-binding sites, 2) not suitable for haptens and molecules not fulfilling 1) above. The advantages for ILSA are as above 1) to 4) for ILMA, in addition to which the use of labelled "second" antibody allows for a "universal" label. Disadvantages are as for ILMA 1) and 2), and the fact that antibody on

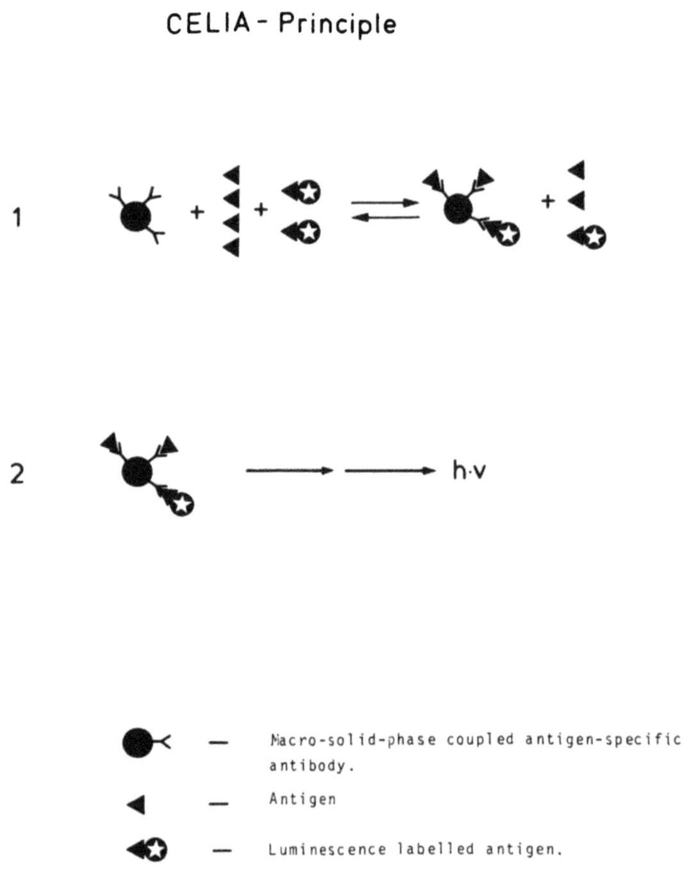

Fig. 6a. CELIA principle. A solid-phase CELIA which is analogous to a solid-phase RIA in which the radioactive label has been replaced by a luminescent one. Between each numbered step is a wash step. This also applies to figs. 6b-6d.

solid-phase and in liquid-phase must be from non-related animal species, e.g. rabbit/sheep. The advantages of the SPALT assay are 1) separation of label and sample incubations, 2) available for small and large molecules, 3) no antigen tracer needed, thus leading to increased sensitivity, 4) labelled "second" antibody

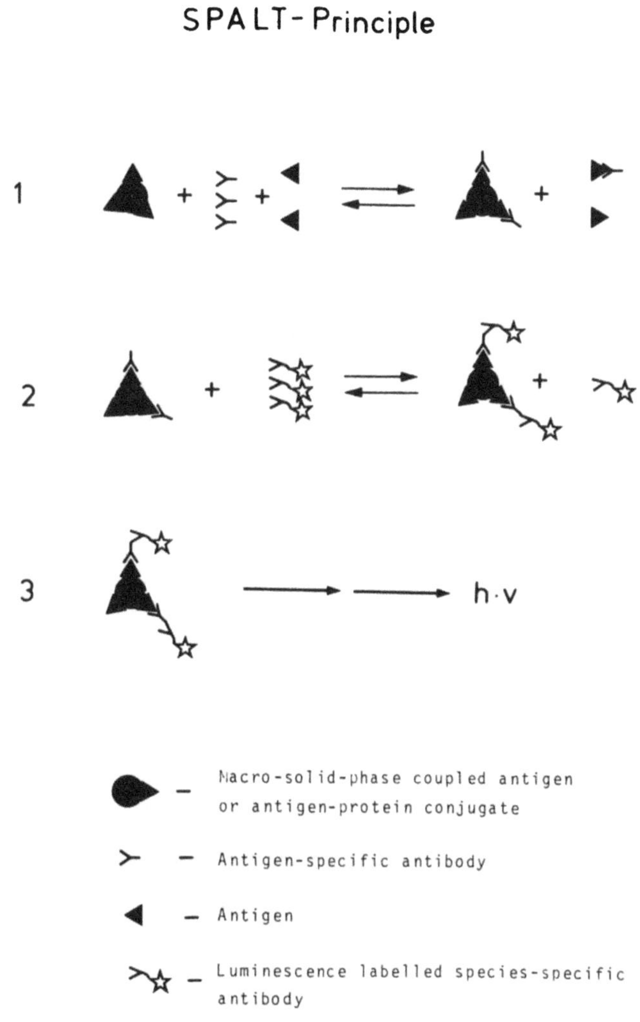

Fig. 6b. SPALT principle. The solid-phase is coated with antigen (or in the case of haptens, antigen-protein conjugate). The assay is analogous to the RAST used in allergy diagnosis. The point to note is that no antigen-tracer is used.

as "universal" label. The single disadvantage is that synthesis of hapten-protein conjugates for solid-phase is critical(43,45-47).

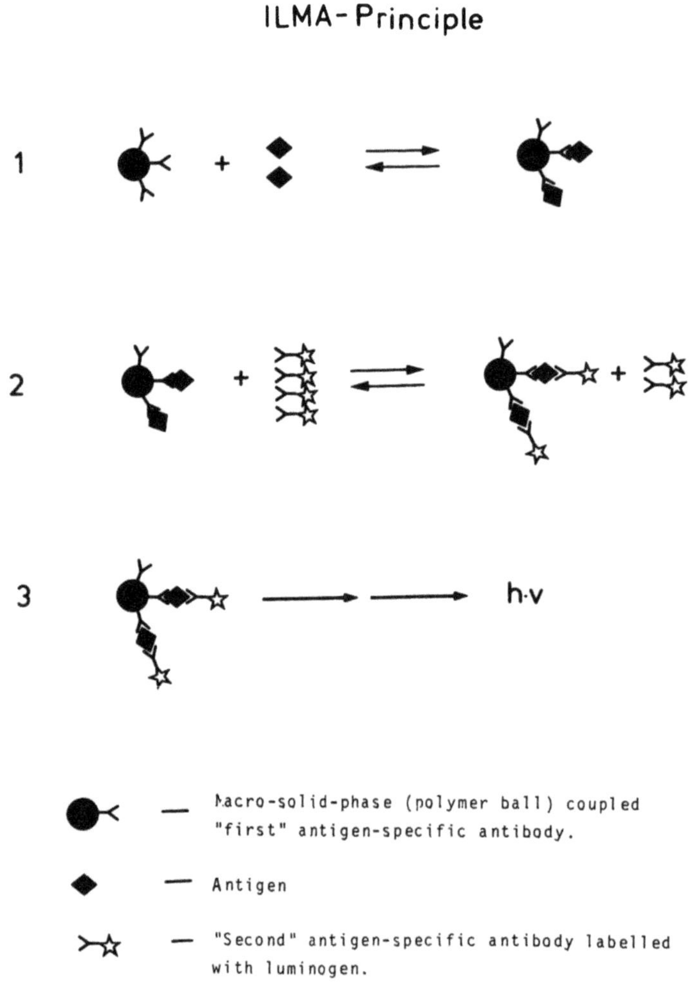

Fig. 6c. ILMA principle. This assay is analogous to the immunoradiometric assay (IRMA).

The advantages and disadvantages of different solid-phases are listed below. Coated tubes: Advantages are 1) Ease of coating using adsorptive techniques known in RIA, 2) simple assay scheme

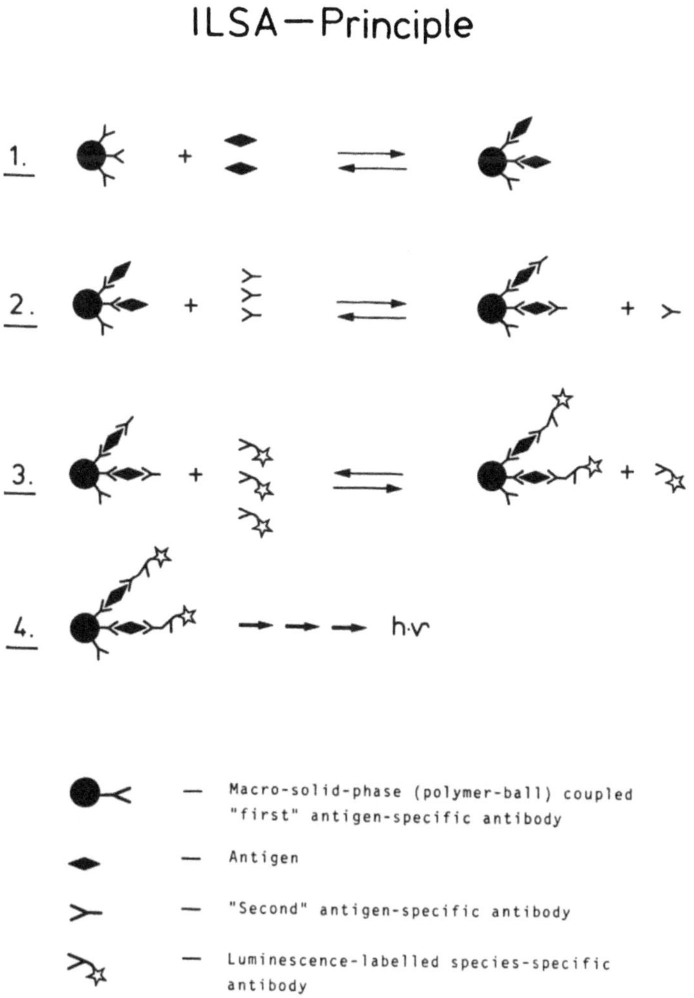

Fig. 6d. ILSA principle. This assay is similar to the ILMA, but the label has been replaced with a species-specific antibody-luminogen conjugate. In SPALT and ILSA assays a "universal" label can be used for all liquid-phase antibodies from one species.

no centrifugation step, can easily be automated. Disadvantages are that 1) tubes must be "luminometer-specific", 2) no tube-change is possible before measurement to reduce non-specific effects, 3) covalent binding of antibodies is "fiddly", 4) large storage volume is needed. Microparticles: Advantages 1) can be dosed accordingly, 2) activation is easy (18, 47), 3) luminometer is "independent", 4) small storage volume. Disadvantages 1) must be kept in suspension - agitation often required, 2) must be centrifuged, 3) pipetting and wash steps time-consuming and difficult to automate, 4) transfer of precipitate to new measuring cuvette impractical. Macroparticles (6.4 mm diameter balls): Advantages 1) easy to dispense, 2) easily automated, 3) no centrifugation, 4) small storage volume, 5) wash systems commercially available, 6) luminometer "independent", 7) tube transfer before measurement easy to perform. Disadvantage 1) they cannot be dosed as easily as micro particles and coating must be optimised for each assay.

There are macro- and micro-solid-phases and, although it is the aim of the author to establish assays using a macro solid-phase in the form of polystyrene balls (6.4 mm diameter), this has not always been possible, and to press, the assays for insulin and insulin antibodies (19) still use microcrystalline cellulose (20 μm particles) as solid-phase.

The CELIA (or EELIA) can only be recommended where a prior extraction step has been performed to remove substances which quench as well as non-specific substances likely to react with the antibody. The problems are similar to those encountered with β-scintillation counting - in both cases, blue light emission (410-430 nm) with absorption if yellow components (e.g. bilirubin) are present. The extraction step must be considered when wishing to set up a CELIA. Possible interference in such assays may occur when using phthalhydrazide compounds (e.g. luminol-derivatives) as labels which may react with haem-components and/or peroxidase present in serum. One way of combatting such reactions is the use of high concentrations of azide in the assay buffer which eliminates the peroxidase, bearing in mind that azide is highly toxic.

The alternative to CELIA for molecules with only one antigenic determinant or epitope is the SPALT, which also works for haptens when set up properly (19, 43, 44). The SPALT is similar in principle to the RAST (radioallergosorbent test - Pharmacia, Uppsala, Sweden). As these assays use immobilised antigens, all the problems of antigen recognition by the antibody come into play, so one must consider carefully the choice of immunogen for the raising of antibodies and that used for immobilisation on the solid phase. These difficulties have been described elsewhere, both for radioimmunoassays (45) and for luminescence immunoassays (46). The reason for such considerations is that the antibodies raised to protein-hapten conjugates do not only react with the hapten, but to the bridge-region (link between hapten and carrier protein) and to the carrier protein itself. Such problems are not seen in RIA or CELIA as a specific antigen-tracer is used which reacts only with the proportion of antibodies specific for the hapten part of the immunogen. As the SPALT uses a species-specific (second) antibody, e.g. ABEI-H labelled donkey-anti-rabbit IgG, all rabbit IgG molecules bound to the solid-phase react with the label. The SPALT method allows, as do both ILMA and ILSA, the necessary separation of analyte and tracer in two separate incubation steps. Although the SPALT assay has one more incubation step than the CELIA, the advantages outweigh this, as substances which interfere with the light-reaction are removed by washing, before the tracer is added.

The ILMA is analogous to the IRMA and can be used in identical situations, the limitation of both assays being that the antigen must have at least two distinct epitopes, which means that a molecular weight in excess of ca 5000 daltons must be present, although there are much larger molecules, for example thyrotropin (TSH) with a molecular weight of around 26,000 daltons, for which it has been difficult to set up IRMA or ILMA methods due to epitope problems.

The ILSA may be used as an alternative to the ILMA where additional sensitivity is required, or where the liquid-phase substance-specific antibody is a monoclonal antibody "sensitive" to labelling, as is the case for TSH-alpha-specific antibody. The disadvantage of the ILSA is that the two substance-specific antibodies must come from different animal-species because of the use of labelled "second" antibody.

Examples of assays published from different laboratories show how the various problems have been tackled. Assays using different labels are also described where data is available.

a. CELIA - Chemiluminescence immunoassays using labelled antigens

The most commonly described ligands to be determined are steroids and their derivatives where a prior extraction step is performed, or where direct measurement is possible, for example in saliva.

i. Liquid-phase homogeneous CELIA (48). Table 2 shows the components and assay scheme for a cortisol CELIA. The extraction from serum/urine is performed with dichloromethane, the antigen being labelled with APEI (6-(5-aminopentyl-N-ethyl)isoluminol) coupled to cortisol 21-hemisuccinate. The reference is good as it describes fully and precisely the validation of the system used, including a comparison with radioimmunoassay.

Table 2. Flow diagram for a homogeneous liquid-phase CELIA(48)

Label: APEI-cortisol-21-hemisuccinate conjugate

Sample extraction solvent: dichloromethane

Assay scheme:	100 µl sample extract / standard
	100 µl anti-cortisol in assay buffer*
	Incubate 20 min / RT^{+}
	100 µl cortisol-APEI
	Incubate 90 min / 4 ^{o}C
	Insert tube in luminometer and inject
	100 µl MP-11 in phosphate buffer, pH 7.4
	100 µl hydrogen peroxide in borate buffer, pH 8.6
	Integrate signal 2-12 sec.

$^{+}$RT=room temperature;*assay buffer-0.05 mol/l phosphate containing 0.11 mol/l NaCl and 20 mg/l bovine serum albumin(BSA),pH 8.0

ii. Solid-phase CELIA (49). Table 3 shows the components and assay flow-diagram of a CELIA for 17β-oestradiol using monoclonal antibodies adsorbed to the reaction vessel (polystyrene tube) wall. The tracer used is ABEI coupled to 17β-oestradiol-6-carboxymethyloxime. The thing to note here is the use of an incubation step with strong NaOH to increase the light output, a practice often used in this type of assay. In contrast to the liquid-phase CELIA described above, the pH at which the light initiation was performed here was strongly alkaline.

Table 3. Assay flow diagram for a solid-phase CELIA (49).

Label: ABEI-oestradiol-6-carboxymethyloxime (CMO) conjugate

Sample extraction solvent: diethyl ether

Assay scheme:	100 µl sample/standard extract
	100 µl oestradiol-ABEI in phosphate buffer
	1 anti-oestradiol (monoclonal) coated tube
	Incubate 60 min / 4°C
	Aspirate tube contents
	Wash with 400 µl phosphate buffer
	300 µl 5mol/l NaOH
	Incubate 30 min / 37°C
	100 µl microperoxidase
	Transfer tube to luminometer
	Inject 100 µl hydrogen peroxide
	Integrate signal over 10 sec.

b. SPALT - solid phase antigen assays using labelled "second" antibody

These assays are as described above, suitable for hapten and protein determinations in non-extracted samples. The method was developed by the author and examples are taken from the routine using a thyroxine SPALT as example for a hapten assay. The assay lay-out and components are shown in Table 4. Table 5 shows the same for thyroxine binding prealbumin or transthyretin (TBPA), this serum component representing the protein SPALT. In both cases the label was ABEI-H, although diazoluminol can also be used (47, 50).

Table 4. Assay for thyroxine - a scheme for a hapten SPALT (43)
Label: Donkey-anti-rabbit serum - ABEI-H (DARS-ABEI-H)
Assay flow diagram:

50 µl sample / standard
100 µl buffer 4L (see Practical Aspects - Assay Buffers)
50 µl rabbit anti-thyroxine
Incubate 10 min / RT
1 Transferrin-Thyroxin conjugate coated ball
Incubate 30 min / RT on horizontal rotator (170-190 rpm)
Wash with saline/Tween (see Practical Aspects - Wash solution)
200 µl DARS-ABEI-H
Incubate 60 min / RT on horizontal rotator
Wash as above, transfer ball to measuring cuvette
Add 300 µl NaCl/microperoxidase
Load luminometer
Inject 300 µl NaOH/urea peroxide
Integrate signal over 20 sec.

Table 5. Assay scheme for an acute-phase protein SPALT (2)
Label: DARS-ABEI-H (see Table 4)
Assay flow diagram:

100 µl sample / serum (1:1000 dilution)
100 µl rabbit-anti transthytretin (TBPA)
Incubate 15 min / RT
1 human-TBPA coated ball
Incubate 45 min / RT on horizontal rotator
Wash as in Table 4
200 µl DARS-ABEI-H
Incubate 60 min / RT on horizontal rotator
Wash and process as for thyroxine SPALT (Table 4)

c. ILMA - solid phase antibody assays using labelled "first" antibody

The term ICMA is also used to stress that the assay uses a chemiluminescent label (ICMA - immunochemiluminometric assay, 34).

The sandwich assays here described are for alpha-foetoprotein using a microcrystalline cellulose solid-phase and acridinium-ester label. This assay type has been successfully developed by the Cardiff group under Drs. Woodhead and Weeks. The second example is a β_1-pregnancy-specific glycoprotein (β_1-SP1) developed for monitoring foetal development after extrauterine fertilisation (2) and using ABEI-H as label. The details are shown in Tables 6 and 7.

Table 6. Assay flow diagram for an alpha foetoprotein ICMA (34)

Label: Anti-alpha foetoprotein acridinium ester conjugate

Assay scheme: 100 µl standard in buffer + 100 µl horse serum

or

100 µl patient serum + 100 µl assay buffer

100 µl anti-AFP solid-phase (MCC - 100 µg)

100 µl anti-AFP-acridinium ester (33 ng)

Incubate 60 min / RT

Wash with 1 ml buffer

Centrifuge 15 min / 2000 g

Aspirate supernatant and repeat wash and centrifugation steps

Add 200 µl distilled water to each tube

Load luminometer

Inject 100 µl NaOH/hydrogen peroxide

Integrate counts over 10 sec.

Table 7. Assay flow diagram for an ILMA developed to control early pregnancy after extra-uterine fertilisation-β_1-SP1

Label: Anti-β_1SP1-ABEI-H

Assay scheme: 50 µl sample / standard

150 µl buffer 4N (see Practical Aspects - Assay buffers)

1 anti β-SP1 coated ball

Incubate 60 min / RT

Wash as in Table 4

200 µl anti-β_1-SP1-ABEI-H

Incubate 60 min / RT

Wash and process as for thyroxine SPALT(Table 4)

d. ILSA - solid-phase antibody assays using labelled "second" antibody

The ILSA described here is one for thyrotropin (TSH) using a combination of monoclonal antibody bound to the solid-phase and a polyclonal rabbit antibody as liquid-phase. Both antibodies can be added in the first incubation, the labelled second antibody being added after a wash step. Table 8 shows the assay flow-sheet and components. The combined first incubation means that the optimised ILSA has, like the ILMA, only two incubation steps. Where the antibody combination does not allow for a combined incubation, three sequential incubation steps are needed, which may present a problem, where time-saving and commercialisation of such assays are contemplated.

Table 8. Scheme for a sensitive thyrotropin (TSH) ILSA (2)

Label: DARS-ABEI-H (see Table 4)

Assay scheme: 200 µl serum / standard*
50 µl rabbit anti-TSH (whole molecule specific)
1 ball - mouse anti-TSH (β-subunit specific)
Incubate 4 hr / RT
Wash as in Table 4
200 µl DARS-ABEI-H
Incubate 2 hr / RT
Wash and process as in Table 4

*WHO-68/38 reference material dissolved in serum from an untreated thyrotoxic patient.

e. EELIA - a solid-phase assay with bioluminescent labelled antigen

The first luminescence immunoassays set up in this laboratory (18) used a bioluminescent labelled antigen, namely pyruvate kinase labelled transferrin, the "second" antibody being immobilised on microcrystalline cellulose. ATP was generated from ADP and phosphoenolpyruvate (PEP), the light being generated from an ATP-monitoring reagent consisting of firefly luciferase-luciferin mixture. ATP was measured kinetically. Table 9 shows

Table 9. A bioluminescent-labelled assay - Transferrin EELIA (18)

Label: Pyruvate kinase-transferrin (PK-Transferrin)

Assay details: 20 µl sample / standards (1:100 dilution)
50 µl rabbit anti-transferrin
50 µl PK-Transferrin
Incubate 45 min / RT
200 µl DARS-MCC*
Incubate 30 min with shaking
Add 1 ml 0.15 mol/l KCl
Mix and centrifuge for 7 min at 3000 g
Repeat wash and centrifuge step
Add following reagent to precipitate
400 µl Tris-HCl / EDTA
100 µl ADP/PEP/$MgCl_2$**
Load tube into luminometer
Inject 100 µl luciferin-luciferase+
Measure ATP production kinetically from the 31st to the 60th second after injection.

*DARS-MCC - Donkey anti-rabbit serum coupled to microcrystalline cellulose particles 20 µm diameter. **ADP - Adenosine diphosphate, PEP - Phospholenol pyruvate. +ATP - Monitoring reagent - LKB-Wallac.

the assay flow-scheme and components for the pioneer transferrin EELIA. The assay was first replaced by a CELIA and then by a SPALT and finally by an ILMA. The EELIA gave results (1%-2%) which could not be explained and which were probably due to inhibition of one of the enzymes (pyruvate kinase or luciferase) during the incubation or measurement step. It is known that the firefly luciferase-luciferin system is susceptible to anion inhibition (51) where thiocyanate and perchlorate inhibit strongly whereas acetate hardly has any effect and is therefore the choice of buffer in this system (52).

It should be noted that the NAD(P)H dependent luciferase-luciferin systems from Photobacterium fischerii and Beneckea harveyi are not inhibited by azide, cyanide and fluoride even at concentrations of 1 mmol/l, but are strongly inhibited by calcium and magnesium ions, even at a concentration of 10 µmol/l (53). The latter can be chelated with EDTA (1 mmol/l) included in the monitoring buffer. The alternative label for bioluminescent dependent systems is a robust dehydrogenase coupled with a bacterial luciferin-luciferase system. All components are available commercially, examples of which are given in Table 10.

Table 10. Some components commercially available as potential bioluminescent labels

a. Kinases (ATP-dependent systems)

Pyruvate kinase - E.C. 2.7.1.40

(Adenylate kinase - E.C. 2.7.4.3) - cave (50)

(Acetate kinase - E.C. 2.7.2.1) (50)

b. Dehydrogenases (NAD(P)H-dependent systems)

Glucose-6-phosphate dehydrogenase - E.C. 1.1.1.49

Glyceraldehyde-3-phosphate dehydrogenase - E.C. 1.2.1.12

Lactate dehydrogenase - E.C. 1.1.1.27

Alcohol dehydrogenase - E.C. 1.1.1.1

c. Oxidases (hydrogen peroxide producing for use with chemiluminescent luminogens)

Glucose oxidase - E.C. 1.1.3.4

Peroxidase - E.C. 1.11.1.7

d. Substrates

DL-Glyceraldehyde-3-phosphoric acid

Glucose-6-phosphate

Lactic acid

Glucose

Adenosine diphosphate

Phosphoenol pyruvate

Acetyl phosphate

e. Co-factors and monitoring reagents

Firefly luciferase

D-Luciferin (synthetic)

NAD

NADP

Table 10 (contd.)
ATP-monitoring reagent (firefly luciferin-luciferase system)
NAD(P)H monitoring reagent (bacterial luciferin-luciferase systems).

f. LEMIT - luminescent modification of the EMIT enzyme immunoassay

In drug monitoring assays where glucose-6-phosphate dehydrogenase (G-6-PDH) is used as label, a modification has been described (54) which increases the sensitivity of EMIT assays using G-6-PDH by a factor of 900 for phenytoin when the photometric determination of NADH is replaced by a bacterial luminescence detection system.

g. LUCIA - labelled co-factor assays with luminescent detection

The use of co-factors as labels has been widely documented, the main labels being derivatives of NAD(P) and ATP (33, 36). Use was made of an NAD-biotin label (33) to measure biotin, using an ethanol/alcohol dehydrogenase/bacterial luciferase-luciferin system. ATP-labelling of 2,4-dinitrobenzene has been described (36) and used in a homogeneous LUCIA for N-(2,4-dinitrophenyl)-β-alanine which had a detection limit of 10 nmol/l (55). One difficulty in the LUCIA is the production of bioactive co-factor ligand conjugates, losses of 75% of the activity being not uncommon as has been described for the coupling of NAD via its 2-amino group to ligands using ethylenimine as coupler-spacer (56). The use of avidin-biotin systems, where one component is labelled with biotin, the other partner being luminogen labelled avidin, is worth mentioning. Such a two-component system is gaining wider use as an alternative labelling scheme for immunoassays (57).

h. Immobilised luminescent reagents - the way to "recycling assays"?

Immobilised luminescent reagents have been produced by several groups of workers (58-60), where components have been adsorbed or covalently coupled to glass or polystyrene balls or "dipsticks". A serious drawback seems to be the loss in activity during the coupling process, immobilisation of firefly luciferase to glass beads using diazotisation resulted in a loss of 99.8% of the activity (60). Better results appear to have been obtained by using arylamine beads (59) and an assay for testosterone has been described using immobilised hydroxysteroid dehydrogenase (61). Whether this type of reagent will play a prominent role in immunoassay is uncertain, but makers of "black-box" automation systems for luminescence immunoassays will no doubt have such reagents in mind!

Instrumentation and Measurement

a. Instrumentation

The lack of availability of quality-controlled reagents as well as specialised yet inflexible instrumentation has thwarted the growth of the technique of analytical luminescence (62). Quality-controlled reagents have become commercially available over the past few years (LKB-Wallac, Turku, Finland; Sigma, St. Louis, USA) making important contributions. The situation as regards instruments has changed very little, there being no instrument entirely suitable for routine luminescence immunoassays, which means an instrument that has full data-processing facilities, can be easily operated and which is reliable enough to prevent loss of samples. The latter is of great importance as the biggest drawback of chemiluminescence is that the measurement cannot be repeated in most cases, in contrast to RIA. Instruments are available which are "half way there", for example having automatic injection possibilities and room for up to 300 samples, but lack the data processing available in most beta- and gamma-counters. The future of luminescence immunoassay as a routine analytical tool hangs in the balance because of the above mentioned shortcomings. An excellent review on instruments is available (62), and manufacturers could be consulted regarding luminometers suitable for immunoassay. Evaluation of luminometers over long periods of time is sparsely documented, with only a few reviews being so far available (46, 62).

b. Measurement

The question as to how the light signal should be generated and measured has been a constant source of discussion during the development of luminescent methodology, not only for immunoassays. The possibilities are: i) peak signal (63), ii) integral measurement of the area under the light output curve (44, 49), iii) integral measurement under part of the light output curve (48, 64), iv) constant light signal (65), v) kinetic measurement (slope) (18, 19), vi) end-point measurement.

For bioluminescent labels, methods ii-v have been used with success; for chemiluminescent labels, methods i-iii. Method vi is included as a possibility which is probably more in keeping with clinical chemical determinations.

The method chosen for measuring the light output is very much dependent on the label and oxidation system. To clarify this, acridinium-ester labels have a short light output with regard to time and integration,over 5 seconds or less is usually sufficient (38). In contrast, luminol-derivatives are measured either at pH 8-9 where the reaction velocity is high, leading to short integration times, or at a pH of 13 or more, where the light output continues over a much longer period of time, integration times of 30 seconds being not uncommon. At strongly alkaline pH, an approximation of method iv can be achieved, together with relatively short measurements of 20 seconds and less, when using phthalhydrazide labels.

When using a bioluminescent system where an enzyme label is introduced to generate ATP or NAD(P)H, the kinetic measurement is the method of choice, with measurements over 10 seconds being possible. In the LUCIA technique, a constant signal or integral measurement can be expected, depending upon the monitoring reagents used (18, 50, 65), the latter being available for both techniques, usually depending upon the sensitivity required.

It must now be apparent that no single optimised system exists, and at best it is to be expected that a small group of methods will become accepted for use in luminescence immunoassays. It should be mentioned that, not only reagent stability is important, but also the speed of injection of reagents, especially at the light generation step, which must remain constant.

Myths and Facts of Luminescence Immunoassays

Many important facts and observations are restricted to the system and reagents used and cannot be transferred to all assays. Other tenets are basic and apply to most, if not all, luminescence immunoassay systems. Information summarised in Table 11 shows that a must for each assay is pure water, but that the hopes of a homologous assay with enhancement are strongly dependent upon the antiserum used. The basics of luminescence immunoassay are the same as for radioimmunoassay. When setting up an LIA for the first time, it is often advisable to consult a laboratory engaged in similar work.

Data Reduction

The problems concerning data processing of luminescence immunoassay are the same as for radioimmunoassay (66). It is to be expected that luminometers with full data processing facilities either as hardware (integrated in EEPROMS) or software (computer programmes) will be available in the near future. Many of the programmes offered will no doubt be direct adaption of existing radioimmunoassay software packets, the changes being in the control of the luminometer itself (injector and photo-

Table 11. Some facts and myths concerning luminescent immunoassays.

Facts - important points to be followed

1. High quality water for reagents imperative (at least quartz distilled)
2. Flushing of luminometer reagent lines before use when the luminometer has stood unused for longer than 5 min
3. Constant speed injection to initiate light reaction
4. Rapid and reliable mixing of reagents in measurement chamber
5. Reliable mechanics for tube change and light exclusion
6. Good geometry for light reflection in measurement chamber
7. Good precision and fine tolerances for cuvettes/test tubes
8. Polystyrene and polypropylene both ideal for cuvettes

Myths - important facts which can be ignored

1. No metal parts in reagent flow-path
2. Temperature control to 0.1^{o}C
3. High-precision special (expensive) cuvettes
4. Continuous mixing during measurement
5. Adjust photon-output to be within the deadtime (ca. 20 ns) of the photomultiplier tube
6. Specially qualified staff to operate luminometer

multiplier control). The same facts apply to luminescence immunoassays as for RIA-automation and computerisation, that is, that the best programmes cannot convert a bad assay into a good one!

Before purchasing a luminometer it is advisable to check that the data processing side is complete, i.e. that all programmes are ready and not "in development" or "available soon". Standard works on RIA data manipulation and to adapt these to LIA are available (66-70).

Quality Control of Reagents and Results

a. Reagents

A limited number of highly purified and well controlled reagents are available. These include luminogens, labelled antigens and antibodies, luciferin-luciferase mixtures (monitoring reagents) as well as accessory reagents (e.g. MP 11 for the oxidation system). Table 10 shows a list of some reagents commercially available, whose quality has been tested over several batches (where possible). The introduction of LIA-kits will no doubt be accompanied by the offer of more suitable quality-controlled reagents.

b. Cuvettes / measuring tubes

The non-standardisation of luminometers has led to the need for specific cuvettes for each machine. Some luminometers allow several tube sizes whereas others need a specific, and often expensive, cuvette size. A standardisation of tube size, for example 75 x 12 mm as in RIA coated tube assays, would be welcomed, especially by kit producers concentrating on the latter technique. The alternative is the use of coated ball assays, which are independent of the luminometer used and which reduce non-specific effects as the ball must be transferred to a clean cuvette before measuring.

The main features of cuvettes, either made of polypropylene or polystyrene, is their precision or tolerance, as most luminometers have been conceived for "multi-use", for example phagocytosis, enzyme measurement and bacterial ATP, luminescence immunoassay being a future possibility. In a series of tubes tested, where the tube size should be 55 x 11 mm, all of four lots from different manufacturers had to be rejected as the tolerance was not satisfactory, some tubes being conical with maximal external diameters in excess of 12.3 mm, others having a lip around the opening, causing the tubes to jam in the measuring chamber.

The cost of cuvettes is an important point to consider when buying a luminometer, especially when the number of assays to be performed is high. It is possible that up to 40% of the assay material costs may be accounted for by "special high precision cuvettes". It may be less expensive in the long run to have a large batch made by a firm specialising in high quality laboratory plastic ware, but the breakeven point in such an adventure may be several million tubes!

c. Antisera

The quality control of antisera is important, especially for hapten assays, be they of the CELIA or SPALT type. In both cases, the choice of hapten-conjugate may play an important role in the binding of the antisera to the hapten-luminogen in the case of the CELIA, and the hapten-protein on the solid phase in the SPALT assay. It is not uncommon to find antisera which are suitable for steroid radioimmunoassay but which perform completely differently in the same steroid CELIA (71). For protein and peptide assays, the same rules for antisera quality apply as for RIA or IRMA with regard to specificity and sensitivity.

d. Standards

Standard materials are available for bioluminescence inasmuch as ATP and NAD(P) are commercially available in pre-weighed amounts. As mentioned above, standardised materials in the form of monitoring reagents are also available. Standard light sources are, as far as is known, not available, the only alternative being a calibrated β-emitter (^{3}H or ^{14}C) in a scintillator solution from a supplier of radiochemicals, although this must be in a vial which can be inserted in the luminometer measurement chamber, which can present problems. The availability of standard light sources is a problem which is being tackled at the present time.

e. External quality assessment

The best way of testing an assay is to participate in a suitable external quality assessment scheme (EQAS), especially in the setting-up phases and clinical trial of a method. EQAS are operated at a national and international level, both centrally and commercially.

f. Data quality control - exclusion of outliers

This process is always difficult, but it is safe to say that the methods for RIA can be transferred to LIA. In addition, light-output curve characteristics can be monitored (e.g. peak height, signal decay rate), and certain laboratories have used such data, together with computer-support to screen for outliers (72) in chemiluminescent immunoassays. The screening of outliers is an important and also debatable point, but it is hoped that luminometer suppliers delivering support programmes will offer such a possibility, based on both practical and theoretical grounds.

Practical Aspects

Practical methods for preparing reagents for luminescence immunoassays are given. All have been tried with success and, although they have been specially designed for use in phthalhydrazide-labelled immunoassays, they should, where relevant, be easy to modify for use with other labels, for example acridinium esters. The topics covered include solid phase activation, coupling of ligands, saturation to reduce non-specific binding, a flexible modular assay system suitable for all luminometers, using commercially available apparatus and preparation of labelled antibodies and antigens. All data refer to the ABEI-H/alkaline peroxide/pseudoperoxidase (MP 11) system used.

a. Solid phase activation

i. Microcrystalline cellulose (MCC). Suitable material is available from Sigma with a particle diameter of 20 µm. Activation is carried out at $50^{o}C$-$70^{o}C$ in 0.1 mol/l phosphate buffer pH5-6 using sodium metaperiodate (ca 10 mg/g MCC) as oxidant. The appearance of aldehyde groups can be monitored using Schiff's reagent, when a lilac-pink coloration is taken to be positive. To carry out the Schiff test, 50 µl slurry is taken, to which 1 ml Schiff's reagent (p-rosaniline decolorised with sulphur dioxide) is added. In the early stages of oxidation, the reaction mixture containing the Schiff's reagent may turn brown due to reduction of periodate to elemental iodine. In such cases, the amount of slurry taken should be reduced.

After the reaction has occurred (1-2 hr) the cellulose is washed copiously with water to remove excess iodate and periodate. This step can be preceded by reduction of the iodate and periodate to iodide using sodium dithionite. The mixture turns brown (iodine) then colourless (iodide). The aldehyde groups are not reduced with dithionite. After washing, the cellulose is suspended in 0.05 mol/l sodium tetraborate which has a pH around 9.35, and an aqueous solution of the ligand rapidly added under stirring (addition time 2-3 s). The coupling reaction is allowed to proceed overnight at $4^{o}C$ after which unreacted aldehyde groups can be saturated with an excess of bovine serum albumin (5 g/l) in 0.05 mol/l Tris-HCl, (pH 8). The saturation reaction is allowed to run overnight at $4^{o}C$ after which, if required, the pH is adjusted to 6-7 and sodium borohydride (10 mg/g MCC) is added and allowed to react for at least 2 hr to stabilise any Schiff's bases present. After this step, the cellulose is washed with physiological saline and distilled water before being dried (lyophilised). In the lyophilised form, the MCC-ligand is stable for several months, suspensions in assay buffer being made up in amounts suitable for 7-14 days use. Examples using MCC-antibodies have already been published (19).

ii. Polystyrene balls. These are an ideal solid phase as they are non-porous (cf. arylamine-resins and agarose-acrylamide supports) and can be obtained if different sizes. A suitable size is 6.4 mm diameter (0.25") and is commercially available

(Precision Plastic Ball Co. Chicago, Illinois, USA., Spherotech Kugeln, Fulda, FRG and Euro-Matic Ltd., Brentford, Middlesex, UK). The amounts of reagents used refer to the 6.4 mm balls.

The easier activation proceeds by first coating the balls with a 3 mg/1000 balls aqueous solution of poly phenylalanine-lysine (1:1 co-polymer) Mr 30-40 kilodaltons, available from Miles-Yeda or Sigma. The coating process can take 24 hr-96 hr either at room temperature or in the refrigerator, the important point being that the balls must be dry before the coating solution is added. After coating, the balls can be dried and stored for several weeks without apparent loss in amino groups or precision.

The terminal amino groups of the lysine residues are activated with a 0.5% solution of purified glutaraldehyde (e.g. Sigma, Cat. No. G-5882) for 30min-45 min after which the balls are washed with physiological saline and distilled water, the presence of aldehyde groups being tested for with Schiff's reagent, using a non-activated ball as control. The "aldehyde" balls must be kept moist or the precision of the ligand-coupling is reduced. An aqueous solution of the ligand (125 ml solution/ 1000 balls) is poured over the activated balls, the vessel gently shaken and allowed to stand for 20 min-30 min, after which the pH is adjusted to between 8.2 and 8.8 with 0.5 mol/l dipotassium hydrogen phosphate (ca 5 ml/1000 balls). The amount of ligand used should have a protein content of around 5-10 mg/1000 balls.

The coupling time is 18 hr-24 hr at room temperature, after which the saturation of non-specific binding sites is effected with bovine serum albumin (5 ml of a 50 g/l solution/1000 balls) for 24 hr followed by addition of 0.5 mol/l Tris-HCl, pH 8 (5 ml/ 1000 balls), after which the mixture is allowed to stand for a further 24 hr. (Both BSA and Tris stock solutions should contain at least 0.1 mol/l sodium azide to prevent bacterial attack during the coating procedure).

After coupling and saturation, the balls are washed with physiological saline and are allowed to stand for 10 min in a 0.25 g/l solution of polyvinyl alcohol (PVA), after which they are dried under a stream of compressed air or nitrogen. Ligand-coated balls prepared in this way are stable for long periods of time, even when stored at room temperature. The PVA-coating, coupled with storage in a dry atmosphere, gives a good resistance to microbial attack. References in which such balls have been prepared and used are available (2, 43, 44).

b. A flexible assay system for use with polystyrene balls

All components are commercially available, the system consisting of the following components: 20/60 well trays (Abbott), Pentawash ^{TM} system (Abbott), Horizontal rotator - 170-200 rpm (Heidolph, Abbott), 20-60 tube boxes (Abbott or do-it-yourself), 51 x 12 mm cuvettes, 55 x 12 mm tubes (Sarstedt). An automatic wash/dispenser is also available (Abbott) for laboratories with a large throughput, which allows for an acceptable level of automation in a routine luminescence immunoassay laboratory at the present time. The use of this system with polystyrene ball assays allows for luminometer-independent assays. The use of the horizontal rotator improves precision and shortens incubation times. Assays using this system have been published (2, 43).

c. Labelling of antigens and antibodies

The starting material for labelling is ABEI-H, (6-(N-(4-amino-butyl-N-ethyl)-isoluminol)-hemisuccinamide) (LKB-Wallac). The alternative is to purchase ABEI (Sigma) and synthesise the hemisuccinamide by reacting ABEI stoichiometrically with succinic anhydride in dry pyridine for 24 hr-72 hr in the dark at room temperature. The ABEI should first be dissolved in the minimal amount of dry dimethyl formamide (DMF). The ABEI-H produced is present under optimal conditions in excess of 80% of the theoretical yield, and can be used without further purification in the following syntheses, except that the pyridine should be removed by vacuum-distillation, the residue being dissolved in dry DMF.

The first step is the synthesis of an ABEI-H active-ester by reacting ABEI-H with N-hydroxysuccinimide (NHS) with an excess of dicyclohexyl carbodiimide (DCCI) in dry DMF overnight at room temperature in the dark. The appearance of needles (2-4 mm long) of dicyclohexyl urea is a good indication that the active ester has been produced. The appearance of an amorphous precipitate indicates that the reagents used were not dry, and that the active ester has not been optimally produced.

At this stage, the active ester can be either used for coupling, or can be portioned and lyophilised for later use. During lyophilisation, the active ester mixture changes colour, but this does not affect its capacity for coupling to ligands. During portioning, the crystals of dicyclohexyl urea are left behind in the reaction vessel. The stability of the lyophilised active ester is in excess of 4 months when stored in the refrigerator.

The ABEI-H-NHS can be coupled to amino group containing ligands (e.g. antigens, antibodies, hapten-derivatives) in an analogous way to the Bolton-Hunter labelling of peptides in radioiodination (73). The coupling takes place at pH 7.8-8.4 in 0.05 mol/l phosphate buffer using a molar excess of the active ester (5-20 fold). The optimal ratio ABEI-H-NHS:ligand must be tried for each ligand, there being no patent recipe as it were. Experiments carried out have already been published (47, 50).

The labelling of the ligand proceeds in the dark at $4^{o}C$-$10^{o}C$ (refrigerator) overnight, after which separation can be carried out using column chromatography with aqueous buffers (proteins and peptides on Sephadex or Ultrogel) or organic eluents (steroid-labels on Sephadex LH-20), or thin-layer chromatography for hapten-labels. As aqueous buffer for peptide elution, 0.05 mol/l Tris-HCl pH 7.5 containing 0.02 mol/l sodium azide has been successfully used, the advantage being that the fractions containing the desired label (maximal specific binding coupled with minimal non-specific binding) can be pooled, portioned and frozen at $-20^{o}C$ for periods in excess of 18 months without apparent loss in binding in the assay system used.

d. Assay buffers for luminescence immunoassays

Although many buffer systems have been described, the following have been used for all chemiluminescence immunoassays developed and carried out in the author's laboratory.

i. PBS-Tween buffer. 0.05 mol/l Phosphate buffer containing 0.15 mol/l NaCl and 0.5 ml/l Tween 20, pH 7.5-8.4 (the pH is antibody dependent).

ii. Buffer 4L/4N. Buffer 4L is a Tris-HCl buffer (0.05 mol/l) containing BSA (2.5 g/l), and 0.1 mol/l KCl, pH 7.5-8.4. Buffer 4N is identical but contains 0.02 mol/l sodium azide. Buffer 4N is ideal for chemiluminescence immunoassays, whereas buffer 4L can be used in bioluminescent systems.

iii. PBS-Tween-4N buffer. This buffer consists of equal portions of buffers i) and ii) above.

iv. Wash solution. 0.1 mol/l NaCl Containing 0.3 ml/l Tween 20 is an ideal wash solution for assays using a covalently-bound antigen/antibody on the solid phase. Ligands which have been purely adsorbed onto the polystyrene surface may very well be removed by this wash solution.

v. Buffer mixtures used in chemiluminescence immunoassays. For most immunoassays using an ABEI-H label, the PBS-Tween-4N buffer has been shown to give good results, especially with regard to low non-specific binding. Labelled antibody or antigen is best stored in buffer 4N in concentrated form, dilutions for use being made with PBS-Tween-4N, which are usually stable as far as binding is concerned, for 3-4 weeks at $4^{\circ}C$, after which reduction in binding is seen, this being due to detergent exposure rather than to microbial attack as the light signal is stable for several weeks.

Future Trends in Immunoassay

If the luminescence immunoassay finds favour commercially as an alternative to radioimmunoassay, especially where sensitive assays are required, then the usual technological developments will be seen, e.g. the development of "black-box" automatic systems suitable for routine use, where reliability and ease of operation are important. In addition, specialised instruments will be developed on a modular basis to cater for all needs together with a simultaneous optimisation of the cost/application ratio.

The luminescence immunoassay will take its place alongside the radioimmunoassay, enzyme immunoassay and fluorescence immunoassay, together with non-label immunoassays such as nephelometry and particle counting immunoassay.

It is impossible to say which assays will survive as commercial interests will to some extent dictate this. The effects of genetic engineering and new advances in *in vitro* immunotechnology will also play an important and hitherto unforseeable role in the fluctuating field of immunoassays.

References

1. Strasburger, C.J., Fricke, H., Gadow, A., Klingler, W., Wood, W.G.: Lumineszenzimmunoassays - Alternativen zum Radioimmunoassay. Ärztl. Lab. 29, 75-82 (1983).*
2. Wood, W.G.: Lumineszenzimmunoassays - ein Jahr voran. Ärztl. Lab. 30, 336-342 (1984).*
3. Kricka, L.J., Whitehead, T.P.: Luminescent immunoassays: new labels for an established technique. Diagnostic Medicine, May Edition, 45-52 (1984).*
4. Whitehead, T.P., Kricka, L.J., Carter, T.J.N., Thorpe, G.H.G.: Analytical luminescence: its potential in the clinical laboratory. Clin. Chem. 25, 1531-1546 (1979).*
5. Kricka, L.J., Carter, T.J.N.: "Clinical and Biochemical Luminescence", Marcel Dekker, New York, Basel, pp. 289 (1982).*

* Reviews

6. DeLuca, M., McElroy, W.D.: "Bioluminescence and Chemiluminescence - Basic Chemistry and Analytical Applications", Academic Press, New York, London, pp. 782 (1981).

7. Johnson, F.H., Haneda, Y.: "Bioluminescence in Progress", Princeton University Press, Princeton, N.J., pp. 650 (1966).

8. Cormier, M.J., Hercules, D.M., Lee, J.: "Chemiluminescence and Bioluminescence", Plenum Press, New York, London, pp. 531 (1973).

9. Barenboim, G.M., Domanskii, A.N., Turoverov, K.K.: "Luminescence of Biopolymers and Cells" (Lyuminestsentsia Biopolimerov i Kletok) Ed. Translation Chen, R.F., Plenum Press, New York, London, pp. 229 (1969).

10. Wilson, T., Schaap, A.P.: The chemiluminescence from cis-diethoxy-1,2-dioxetane - an unexpected effect of oxygen. J. Amer. chem. Soc. 93, 4126-4136 (1971).

11. Wilson, T.: Chemiluminescence in the liquid phase: termal cleavage of dioxetanes. In: Int. Rev. Phys. Chem. Ser. 2, vol. 9, "Chemical Kinetics", Ed. Herschbach, D.R., Butterworths, London, Boston, pp. 265-322 (1976).

12. Boyle, R.: Observations and trials about the resemblances and differences between a burning coal and shining wood. Phil. Trans. Roy. Soc. (Lond) 2, 605-612 (1667).

13. Boyle, R.: Some observations about shining flesh. Phil. Trans. Roy. Soc. (Lond) 7, 5108-5116 (1672).

14. Dubois, R.: Note sur la physiologie des pyrophores. Comp. Rend. Soc. Biol. 37, 559-562 (1885).

15. Dubois, R.: Fonction photogénique chez le Pholas dactylus. Comp. Rend. Soc. Biol. 39, 564-565 (1887).

16. McElroy, W.D.: The energy source for bioluminescence in an isolated system. Proc. natn. Acad. Sci. U.S.A. 33, 342-345 (1947).

17. McElroy, W.D., Seliger, H.H., White, E.H.: Mechanism of bioluminescence, chemoluminescence and enzyme function on the oxidation of firefly luciferin. Photochem. Photobiol. 10, 153-170 (1969).

18. Fricke, H., Strasburger, C.J., Wood, W.G.: Enzyme enhanced luminescence immunoassay for the determination of transferrin in serum. J. Clin. Chem. Clin. Biochem. 20, 91-94 (1982).

19. Wood, W.G., Fricke, H., von Klitzing, L., Strasburger, C.J., Scriba, P.C.: Solid phase antigen luminescence immunoassays (SPALT) for the determination of insulin, insulin antibodies and gentamicin levels in human serum. J. Clin. Chem. Clin. Biochem. 20, 825-831 (1982).

20. Radziszewski, B.: Untersuchungen über Hydrobenzamid, Amarin und Lophin. Ber. Chem. Ges. 10, 70-75 (1877).

21. Albrecht, H.O.: Über die Chemiluminescenz des Aminophthalsäurehydrazids. Z. Phys. Chem. 136, 321-330 (1928).

22. Gleu, K., Petsch, W.: Die Chemiluminescenz der Dimethyldiaridyliumsalze. Angew. Chem. 48, 57-59 (1935).

23. Hersch, L.S., Vann, W.F., Wilhelm, S.A.: A luminol-assisted competitive binding immunoassay of human immunoglobulin G. Anal. Biochem. 93, 267-271 (1979).

24. Kohen, F., Kim, J.B., Barnard, G., Lindner, H.R.: An assay for urinary estriol-16-alpha-glucuronide based on antibody enhanced chemiluminescence. Steroids 36, 405-419 (1980).

25. Weeks, I., Beheshti, I., McCapra, F., Campbell, A.K., Woodhead, J.S.: Acridinium esters as high-specific-activity labels in immunoassay. Clin. Chem. 29, 1474-1479 (1983).

26. McCapra, F., Tutt, D.E., Topping, R.M.: British Patent No. 1,461,877 (1977).

27. Shapiro, R., Chan, J., Pierson, A., Vaccaro, K., Quick, J.: Protein enhanced fluorescein chemiluminescence used in an immunoassay for rubella antibody in serum. Clin. Chem. 30, 889-893 (1984).

28. Cyalume Light Stick - American Cyanamid Company, Stamford, Connecticut, U.S.A.

29. Rauhut, M.M.: Chemiluminescence. In: "Kirk-Othmar Encyclopaedia of Chemical Technology" 3rd edition, Eds. Grayson, M., Echroth, D., Wiley, New York, vol. 5, pp. 416-450 (1979).

30. Suzuki, N., Tsukamoto, T., Izawa, Y.: Chemiluminescence of N-methyl-9-(dicarboalkoxymethyl) acridanes: successive formation of two 1,2-dioxetanone intermediates. Tetrahedron Letters 24, 3005-3008 (1983).

31. Brolin, S.E., Borglund, E., Tegner, L., Wettermark, G.: Photokinetic microassay based on dehydrogenase reactions and bacterial luciferase. Anal. Biochem. 42, 124-135 (1971).

32. Tsuji, A., Maeda, M., Arakawa, H.: Enzyme immunoassay monitored by chemiluminescence reaction using bis-(2,4,6-trichlorophenyl) oxalate fluorescent dye. In: "Proceedings of the 3rd International Symposium on Analytical Applications of Bioluminescence and Chemiluminescence", Birmingham, U.K. Eds. Kricka, L.J., Whitehead, T.P., Academic Press, pp. 253-256 (1984).

33. Schroeder, H.R., Vogelhut, P.O., Carrico, R.J., Boguslaski, R.C., Buckler, R.T.: Competitive protein binding assay for biotin monitored by chemiluminescence. Anal. Chem. 48, 1933-1937 (1976).

34. Weeks, I., Campbell, A.K., Woodhead, J.S.: Two-site immunochemiluminometric assay for human alpha-1 foetoprotein. Clin. Chem. 29, 1480-1483 (1983).

35. Wood, W.G.: Unpublished data.

36. Carrico, R.J., Yeung, K.K., Schroeder, R.C., Buckler, R.T., Christner, J.E.: Specific protein-binding reactions monitored with ligand-ATP conjugates and firefly luciferase. Anal. Biochem. 76, 95-110 (1976).

37. Arakawa, H., Maeda, M., Tsuji, A.: Enzyme immunoassay of cortisol by chemiluminescence reaction of luminol-peroxidase. Bunsaki Kagaku 26, 322-326 (1977).

38. Weeks, I., Sturgess, M., Siddle, K., Jones, M.K., Woodhead, J.S.: A high sensitivity immunochemiluminometric assay for human thyrotropin. Clin. Endocrinol. 20, 489-495 (1984).

39. Williams, D., Huft, G.F., Seitz, R.: Evaluation of peroxyoxalate for determination of enzyme generated peroxide. Anal. Chem. 48, 1003-1006 (1976).

40. Rauhut, M.M., Bollyky, J., Roberts, B.G., Loy, M., Whitman, R.H., Iannotta, A.V., Semsel, A.M., Clarke, R.M.: Chemiluminescence from reactions of electronegatively substituted aryl oxalates with hydrogen peroxide and fluorescent compounds. J. Amer. chem. Soc. 89, 6515-6521 (1967).

41. Whitehead, T.P., Thorpe, G.H.G., Carter, T.J.N., Groucott, C., Kricka, L.J.: Enhanced luminescence procedure for sensitive determination of peroxidase-labelled conjugates in immunoassay. Nature 305, 158-159 (1983).

42. Wood, W.G., Hantke, U., Gross, A.J.: Initiation of luminol-based luminescence by the injection of a single reagent coupled with an enhancement of light output. J. Clin. Chem. Clin. Biochem. 23, 47-49 (1985).

43. Gadow, A., Wood, W.G., Scriba, P.C.: Lumineszenz-Immunoassays für die Bestimmung von Schilddrüsen-parametern - eine Alternative zum Radioimmunoassay. Akt. Endokr. Stoffw. 5, 13-21 (1984).

44. Wood, W.G., Fricke, H., Haritz, J., Krausz, H-S., Tode, B., Strasburger, C.J., Scriba, P.C.: An evaluation of four different luminescence immunoassay systems: CELIA (chemiluminescent immunoassay), SPALT (solid-phase antigen luminescence technique), ILMA (immunoluminometric assay) and ILSA (immunoluminometric labelled second antibody), Part III of a critical study of macro solid phases for use in immunoassay systems. J. Clin. Chem. Clin. Biochem. 22, 349-356 (1984).

45. Hunter, W.M.: Recent advances in radioimmunoassay and related procedures. In: "Radioimmunoassay and Related Procedures in Medicine 1982", International Atomic Energy Authority, Vienna, pp. 3-21 (1982).

46. Wood, W.G.: Luminescence immunoassays - problems and possibilities. J. Clin. Chem. Clin. Biochem. 22, 905-918 (1984).

47. Wood, W.G., Gadow, A.: Immobilisation of antibodies and antigens on macro solid phases - a comparison between adsorptive and covalent binding - Part 1 of a critical study of macro

solid phases for use in immunoassay systems. J. Clin. Chem. Clin. Biochem. 21, 789-797 (1983).

48. Kohen, F., Pazzagli, M., Kim, J.B., Lindner, H.R.: An immunoassay for plasma cortisol based on chemiluminescence. Steroids 36, 421-437 (1980).

49. Kim, J.B., Barnard, G.J., Collins, W.P., Kohen, F., Lindner, H.R., Eshhol, Z.: Measurement of plasma estradiol-17β by solid-phase chemiluminescence immunoassay. Clin. Chem. 28, 1120-1124 (1982).

50. Gadow, A., Fricke, H., Strasburger, C.J., Wood, W.G.: Synthesis and evaluation of luminescent tracers and hapten-protein conjugates for use in luminescence immunoassays with immobilised antibodies and antigens. Part II of a critical study of macro solid phases for use in immunoassay systems. Clin. Chem. Clin. Biochem. 22, 337-347 (1984).

51. Denburg, J.L., McElroy, W.D.: Anion inhibition of firefly luciferase. Anal. Biochem. Biophys. 141, 668-675 (1970).

52. Gilles, R., Pequeux, A., Saive, J.S., Spronck, A.C., Thome-Lentz, G.: Effect of various ions on ATP-determinations using the "luciferine-luciferase" system. Arch. int. Physiol. Biochim. 84, 807-817 (1976).

53. Stanley, P.E.: Determination of subpicomole levels of NADH and FMN using bacterial luciferase and the liquid scintillation spectrometer. Anal. Biochem. 39, 441-453 (1971).

54. Stanley, P.E.: Quantitation of picomole amounts of NADH, NADPH and FMN using bacterial luciferase. Meth. Enzymol. 57, 215-222 (1978).

55. Carrico, R.J., Johnson, R.D., Boguslaski, R.C.: ATP-labelled ligands and firefly luciferase for monitoring protein binding reactions. Meth. Enzymol. 57, 113-122 (1978).

56. Kohen, F., Hollander, Z., Yeager, F.M., Carrico, R.J., Boguslaski, R.C.: A homogeneous enzyme immunoassay for estriol monitored by co-enzymic cycling reactions. In: "Enzyme Labelled Immunoassay of Hormones and Drugs", Ed. Pal, S.B., Walter de Gruyter, Berlin, New York, pp. 67-79 (1978).

57. Leipold, B., Remy, W.: Use of avidin-biotin-peroxidase complex for measurement of UV lesions in human DNA by micro-ELISA. J. Immunol. Meth. 66, 227-234 (1984).

58. Haggerty, C., Jablonski, E., Stav, L., DeLuca, M.: Continuous monitoring of reactions that produce NADH and NADPH using immobilised luciferase and oxidoreductases from Beneckea harveyi. Anal. Biochem. 88, 162-173 (1978).

59. Jablonski, E., DeLuca, M.: Immobilisation of bacterial luciferase and FMN reductase on glass rods. Proc. natn. Acad. Sci. U.S.A. 73, 3848-3851 (1976).

60. Lee, Y., Jablonski, E., DeLuca, M.: Immobilisation of firefly luciferase on glass rods: properties of the immobilised enzyme. Anal. Biochem. 80, 496-501 (1977).

61. Jablonski, E., DeLuca, M.: Properties and uses of immobilised light-emitting systems from Beneckea harveyi. Clin. Chem. 25, 1622-1627 (1979).

62. Stanley, P.E.: Instrumentation. In: "Clinical and Biochemical Luminescence", Eds. Kricka, L.J., Carter, T.J.N., Marcel Dekker, New York, Basel, pp. 219-260 (1982).

63. Schroeder, H.R., Yeager, F.M., Boguslaski, R.C., Vogelhut, P.O.: Immunoassay for serum thyroxine monitored by chemiluminescence. J. Immunol. Meth. 25, 275-282 (1979).

64. Pazzagli, M., Kim, J.B., Bolelli, G-F., Martinazzo, G., Salerno, R.: Luminescent immunoassay (LIA) for steroids. In: "Monoclonal Antibodies and Developments in Immunoassay", Eds. Albertini, A., Ekins, R.P., Elsevier North-Holland, Amsterdam, pp. 147-156 (1981).

65. Webster, J.J., Chang, J.C., Howard, J.L., Leach, F.R.: A comparison of commercial firefly luciferases. In: "Bioluminescence and Chemiluminescence - Basic Chemistry and Analytical Applications", Eds. DeLuca, M., McElroy, W.D., Academic Press, New York, London, pp. 755-762 (1981).

66. Wood, W.G., Sokolowski, G.: Radioimmunoassay in Theory and Practice - A Laboratory Handbook. Schnetztor, Konstanz, pp. 233 (1982).

67. Ekins, R.P.: Automation of radioimmunoassay and other saturation procedures. In: "Radioimmunoassay and Related Procedures in Medicine", International Atomic Energy Authority, Vienna, vol. 1, pp. 91-109 (1974).

68. Healy, M.J.R.: Statistical analysis of radioimmunoassay data. Biochem. J. 130, 207-210 (1972).

69. Marschner, I., Erhardt, F.W., Scriba, P.C.: Calculation of the radioimmunoassay standard curve by spline function. In: "Radioimmunoassay and Related Procedures in Medicine", International Atomic Energy Authority, Vienna, vol. 1, pp. 111-121 (1974).

70. Rodbard, D.: Statistical aspects of radioimmunoassays. In: "Principles of Competitive Protein Binding Assays", Eds. Odell, W.D., Daughaday, W.H., Lippincott, Philadelphia, Toronto, pp. 204-259 (1971).

71. Barnard, G., Collins, W.P., Kohen, F., Lindner, H.R.: The measurement of urinary estriol-16-alpha-glucuronide by a solid-phase chemiluminescent immunoassay. J. Steroid Biochem. 14, 941-948 (1981).

72. Pazzagli, M., Tommasi, A., Damiani, M., Magini, A., Serio, M.: On-line computer analysis of chemiluminescent reactions and application to luminescent immunoassays. In: "Proceedings of

the Third International Symposium on Analytical Applications of Bioluminescence and Chemiluminescence", Birmingham, U.K., Eds. Kricka, L.J., Whitehead, T.P., Academic Press, New York, London, pp. 469-474 (1984).

73. Bolton, A.E., Hunter, W.M.: The labelling of proteins to high specific radioactivities by conjugation to a ^{125}I-containing acylating reagent. Biochem. J. 133, 529-539 (1973).

NON-ISOTOPIC IMMUNOASSAYS FOR THE ESTIMATION OF STEROID HORMONES

U. M. Joshi
Institute for Research in Reproduction (ICMR), Parel, Bombay: 400 012, India

I. Introduction

Competitive immunoassays for the estimation of steroid hormones using labels other than radioisotopes are slowly gaining acceptance and popularity because of several advantages that these techniques have over the widely used radioimmunoassays (RIA). The labels used in non-isotopic immunoassays (NIIA) are non-hazardous to handle and have a long shelf life. Federal rules are stringent for the use of radioisotopes whereas non-isotopic labels require only routine laboratory care. Moreover, unlike radioisotopes, there is no waste disposal problem with these labels.

The two most widely used labels in NIIA of steroids are the enzymes and the chemiluminescent molecules, although assays using fluorescent labels have also been described.The steps involved in the enzyme immunoassays (EIA), chemiluminescence assays (CIA) and fluoroimmunoassays (FIA) for steroids are similar to those for RIA, i.e. competition of labelled material with unlabelled material for binding to antibody, separation of antibody-bound label from unbound label and final quantitation of the signal received either from the bound or from the free fraction. Some assays not requiring the separation of bound from free (homogeneous assays) have also been described.

Immunoassay Technology, Volume 1

II. Enzyme-Immunoassays (EIA)

In 1971 it was suggested (1, 2) that enzymes could be used as labels in the quantitation of nanomolar concentrations of macromolecules by immunoassay procedures. Since then, enzyme immunoassays (EIA) have been described for a number of substances including steroids. It must be remembered however that, compared to the macromolecules in biological fluids, steroids have certain features that add a new dimension to the immunoassay methodology. Firstly, steroids are bound to plasma proteins and therefore they need to be extracted with a solvent before the immunoassay step. The solvent used for extraction should be highly efficient because internal recovery assessment cannot be made using enzyme labelled steroids, which is possible with radioactively labelled steroids. The solvent extraction step can however be avoided by developing direct nonextraction type of methods. Secondly, steroids that are structurally similar, cross-react to a certain extent with the antibody. The cross-reactivity of substances present in relatively high concentration can be avoided by introducing a purification step such as solvent partition and/or chromatography. However, procedural losses are imminent and for reasons already mentioned, enzyme-labelled steroids cannot be used for monitoring these losses. This problem can however be overcome by using highly specific antibodies in steroid EIA.

A. Enzymes used in steroid EIAs

Horseradish peroxidase (E.C.1.11.1.7) and β-galactosidase (E.C. 3.2.1.23) are the most commonly used enzymes in steroid EIAs, although alkaline phosphatase (E.C.3.1.3.1) glucose-6-phosphate dehydrogenase (E.C.1.1.1.49) and glucoamylase (E.C.3.2.1.3) have also been tried. This laboratory has developed EIAs for both the protein and the steroid hormones using the enzyme penicillinase (E.C.3.5.2.6). The choice of a particular enzyme for development of EIA for steroids is governed by a number of factors such as: a) turn over number of enzyme (μ moles of substrate converted per minute by one μ mole of enzyme, b) availability and cost of enzyme and its substrate, c) retention of enzyme activity in the presence of solvents and reagents used for conjugation of steroid to enzyme, d) method used for determination of enzymatic activity, e) kind of instruments available for the measurement of optical density, f) in the case of a homogeneous enzyme immunoassay which does not require separation of antibody bound hormone from unbound, it is necessary to ensure that the enzyme used for labelling of steroid is not present in the test material.

Table 1 gives relevant data on enzymes that have been used to-date for steroid EIA.

As can be seen from the table, estimation of glucose-6-phosphate dehydrogenase and glucoamylase require a U.V. spectrophotometer with attachments for kinetic measurements, while the rest of the enzymes can be estimated using a simple visible range spectrophotometer. In most of the published assays using horseradish peroxidase as a marker, O-phenelene diamine is used as a substrate which is photosensitive and is also considered to be mutagenic; β-galactosidase is an enzyme with high molecular weight. For carrying out the conjugation reaction with a certain molar ratio of steroid to enzyme, the mass of this enzyme in the reaction will be the highest as compared to other enzymes. Besides this, the relative cost of the enzyme and its substrate is higher than for the other enzymes. This laboratory has used enzyme penicillinase because of several advantages such as lower molecular weight (29,000), stability at ambient temperature, ease of estimation, low cost of substrate and enzyme, and absence from biological fluids.

B. Methods for conjugation of steroids with enzymes

Methods used for preparation of steroid conjugation to enzymes and for steroid immunogens for producing antisera are essentially similar. The three most commonly used methods are the mixed

TABLE 1. ENZYMES USED IN ENZYME-IMMUNOASSAY (EIA) OF STEROIDS

Enzyme	Commonly used substrate in EIA of hormones	Method of estimation	Instrument required	Molecular wt.	Turn-over* number	Prescribed optimum temp.
HRP	O-phenelene diamine+H_2O_2	end point	Spectrophoto meter	40000	216000	25°C
Alkaline Phosphatase	p.nitrophenyl phosphate	end point	"	100000	100000	37°C
β-galactosidase	O-nitrophenyl β-D galacto-pyranoside	end point	"	540000	486000	37°C
Glucose-6-phosphate dehydrogenase	Glucose-6-phosphate	kinetic	UV spectro-photometer with attachments for kinetics measurement	102000	-	25°C
Gluco-amylase	Glucose	kinetic		-	-	55°C
Penicillinase	Penicillin	end point	Spectrophoto meter	29000	58000	25°C

* Turn over number is μ mole substrate converted per minute by 1 μ mole of enzyme at prescribed temperature for assay.

anhydride method (3), carbodiimide method (4) and a modified carbodiimide method using an activated steroid ester prepared with N-hydroxy-succinimide (5).

Most of the steroids are hydrophobic and have to be dissolved in an organic solvent. Enzymes, on the other hand, may be destroyed by these solvents, therefore, the amount of solvent used in the reaction should be optimised. The site of the steroid for conjugation should be chosen so that the antibody recognising sites do not get blocked. Since the methods of steroid-enzyme conjugate preparation and immunogen preparation are similar, it is possible that, sometimes, the 'bridge' is recognised by the antibody much more strongly than the steroid itself.

The conjugates prepared by the methods described above are usually purified either by dialysis or by passing them over a small Sephadex^R G25 column. The unreacted steroid, but not the unreacted enzyme, can thus be removed from the conjugate. Only one group of workers have tried to remove the unreacted enzyme by affinity chromatography (6). However, the method described is quite laborious and the loss of enzyme and steroid, though not indicated, are likely to be substantial.

C. Methods for studying efficiency of conjugation

1. Methods for recovery of enzyme. If the enzyme protein has a specific absorption maximum, the concentration of enzyme protein can be ascertained by spectrophotometry. Alternatively, protein estimates can be obtained by following absorption at 280 nm. A sensitive method for protein estimation such as Lowry's or biuret reaction may also be employed. However, these methods will give no idea as to whether or not the recovered protein has also retained its enzymatic activity.

The enzymatic activity of the conjugate can be determined using routine methods. However, it was found (7) that the optimum conditions of pH, substrate concentration, activators used etc. for the conjugated enzyme and native enzyme, differed.

The characteristics of the enzyme in the conjugate may also be studied by determining the Km and Vmax of the conjugate (8).

It is possible that the entire enzymatic activity present in the conjugate is not in a hapten-conjugate form. Part of the enzyme may be free. For the purpose of EIA, the activity associated with hapten (immunoreactivity) is important. This can be determined by finding out enzyme activity in the conjugate bound to excess antibody. The antibody bound form can be separated using solid phase antibody or a double antibody.

2. Methods for recovery of steroid. This can be determined in three ways. Firstly, a radioactive steroid derivative may be introduced along with unlabelled compound during the process of conjugation. The amount of radioactivity incorporated in the conjugate can give some idea about the moles of steroid incorporated per mole of enzyme. The disadvantage of this method is that even the steroid which is sterically hindered by the enzyme, and hence not recognised by the antibody, is measured. The incorporation may also be judged by the classical procedure of determining U.V. absorption spectrum of the standard steroid and of the enzyme, and comparing it with that of the conjugate. Here it is assumed that the characteristic U.V. absorption spectrum of steroid and enzyme remains unaltered even after conjugation, an assumption which need not necessarily be true.

Lastly, the steroid present in the conjugate can be estimated in a RIA system. Here, only the steroid sterically not hindered by the enzyme, and hence available for immunoreaction, is recognised by the antibody and measured. Ideally, since the conjugate is to be used in an immunoassay system, this way of estimating steroid in the conjugate is the most appropriate. Problems could arise, however, if the bridge in the steroid-enzyme conjugate is recognised much more strongly than the steroid itself, resulting in gross overestimation in terms of steroid incorporation. One such overestimation was observed in this laboratory (unpublished data) in a case of conjugation of pregnanediol glucuronide to penicillinase.

D. Factors affecting efficiency of conjugate preparation

The efficiency of conjugation represents the amount of immunoreactive enzyme recovered. Factors that will determine the efficiency are: 1) type of reaction used for conjugation, 2) proportion of reactants, i.e. enzyme and steroid, and 3) heterogeneity/homogeneity of labelling sites for preparation of immunogen and of enzyme conjugate.

1. Influence of type of reaction used for conjugation on efficiency of conjugation. There have been very few studies which have looked into this aspect critically. The recovery of steroid and enzyme in relation to the process of conjugation has been studied (9). Four methods of conjugation, i.e. the mixed anhydride method (MA), carbodiimide method using water soluble carbodiimide (EDG) or using water insoluble carbodiimide (CMC) or using activated ester of steroid with N-hydroxy succinimide and dicyclohexyl carbodiimide (H/DCC), were used to conjugate three steroids: estrone-3β glucuronide (E_1G), estriol-3β glucuronide (E_3G) and pregnanediol glucuronide (PdG) to four enzymes: glucose-6-phosphate dehydrogenase (G_6PD), urease, horseradish peroxidase (HRP) and alcohol dehydrogenase (ADH). It was concluded that the mixed anhydride method was the best for conjugation since it gave maximum recovery of steroid and enzyme. Best results were obtained with HRP, the recovery of enzyme protein and of specific activity of the enzyme being 65%-75%. In the case of other enzymes, recovery ranged from 15% to 45%.

In the preparation of conjugates of progesterone (P) with enzymes, three types of reaction were used. Thus, when P was conjugated with HRP by MA reaction (10), the recovery of enzyme was 15%-20%. When carbodiimide reaction was used to conjugate β-galactosidase to P, 30% enzyme activity was recovered (11). However, when an active ester of P with N-hydroxy succinimide was prepared for carbodiimide reaction with β-galactosidase, the enzyme recovery was found to be 100% (12).

When testosterone was conjugated to HRP by the mixed

anhydride method (7), recovery of steroid was 8%-10%, that of enzyme protein 38%, and of specific activity of enzyme 77%, and recovery of immunoreactive enzyme 27%. The carbodiimide method has been tried in this laboratory for conjugation of testosterone, estrone glucuronide and pregnanediol glucuronide to penicillinase, when recovery of steroid, enzyme activity and immunoreactivity of enzyme were shown to be 0.8%-1.2%, 30%-50% and 5.6%, respectively. It is thus clear that there is no evidence for a particular reaction being more efficient for conjugation than the other.

2. Influence of the proportion of reactants (steroid and enzyme) on the quality of conjugate. In general, it is observed that the molar ratio of steroid to enzyme in the conjugate will be proportional to the molar ratio during reaction. However, as discussed before, each method for studying the incorporation of steroid in the conjugate has certain disadvantages.

It has been suggested (13) that the sensitivity of the assay is better if the molar incorporation ratio of steroid per mole of enzyme is low, an observation corroborated by others (Table 2). However, at least one report has shown (14) that, to obtain a good EIA for cortisol and β-methasone, the number of steroid molecules incorporated per molecule of β-galactosidase has to be more than 20. Conjugates with less molar incorporation ratios were less immunoreactive.

3. Influence of site on steroid for labelling with enzyme, on the sensitivity of EIA. In any steroid molecule, a number of sites are available for labelling with protein or enzyme for the preparation of immunogen or of the conjugate. When the same site is used for the preparation of both, the combination is called homologous, whereas when different sites are chosen for preparations, the combination is termed heterologous.

For the ELISA and DHEA, three conjugates were prepared with HRP, i.e. at 11th, 3rd and 17th position of DHEA, and were tested against two antisera to DHEA-11 BSA and DHEA-3 BSA respectively (15). It was found that the use of a heterologous

TABLE 2. EFFECT OF STEROID TO ENZYME RATIO IN THE CONJUGATE ON THE SENSITIVITY OF EIA

Steroid	Enzyme	Steroid: enzyme ratio	Separation step	Sensitivity	Ref.
Testosterone	HRP	6.6	PEG	12 pg/tube	23
Progesterone	HRP	6.5	Seph	93 pg/ml	24
Cortisol	Alkaline phosphatase	12	DA	1 ng/tube	25
Cortisol	β-galactosidase	4	DA	100 pg/tube	8
			DASP	150 pg/tube	
Testosterone	HRP	1.2	Ab cellulose	4 pg/tube	26
β-methasone	β-galactosidase	12.5 -25	DA	0.15 pg/tube	14
Cortisol	β-galactosidase	20-30	DA	10 pg/tube	27
Testosterone	Penicillinase	1	DA	12 pg/tube	28

PEG : polyethylene glycol   Seph : Antibody coated on sepharose

DA : double antibody   DASP : double antibody solid phase

Ab cellulose : Antibody coated on microcrystalline cellulose.

combination led to a more sensitive assay.

However, it was observed (16) that a homologous combination for E_3 (Antibody to E_3-16/17 monosuccinyl BSA and label E_3-16/17 monosuccinyl-HRP) gave a more sensitive assay than the heterologous combination.

In the case of progesterone, it was reported (10) that the conjugates 11-hydroxy progesterone succinyl HRP and progesterone-3-oxime-HRP displaced nonradioactive progesterone from antibody to 11-hydroxy progesterone succinyl BSA to the same extent.

There are at least two reports which suggest that a heterologous combination may not be suitable for EIA. It has been shown (17) that, in a system using P-11-hydroxy hemisuccinate-β-galactosidase as label and antibody to P-3 CMO BSA, there was no inhibition of binding by unlabelled P, whereas homologous combination of P-11-HS β-galactosidase and antibody to P-11-HS BSA gave a sensitive assay. Similarly, it was observed (18) that a conjugate of cortisol 21 hemisuccinate-alkaline phosphatase showed immunoreactivity to the antibody to cortisol-21 HS BSA but not to the antibody to cortisol-3 oxime BSA.

E. Storage and stability of conjugates

It is a general observation that the steroid-enzyme conjugates have a long shelf life when preserved at $4^{o}C$-$5^{o}C$. They should not be stored below $0^{o}C$ since repeated freezing and thawing would destroy the enzymatic activity. There are, however, very few published studies on systematic evaluation of stability of conjugates.

It has been reported (7) that the conjugate of testosterone with HRP when diluted to a concentration of 930 ng/ml, is stable at $5^{o}C$ over a month. At a concentration of 93 ng/ml it is stable for 6 days, while further dilutions, if needed, have to be made immediately before use. A volume of 10 μl of the conjugate containing 930 ng/ml in phosphate buffered saline, when lyophilised and stored at $-15^{o}C$, is stable for at least a week.

Stock conjugates in this laboratory are preserved with the addition of 1% BSA and 0.1% sodium azide, at $5^{o}C$. The stock conjugate of testosterone penicillinase was found to be stable for over three years. A ten times diluted conjugate is stable for 3 to 4 months at $5^{o}C$.

F. Cross-reactivities with other steroids in EIA as compared to those in RIA

Since the antiserum used for comparison of the two systems is the same, it is expected that cross-reactivities with other steroids in EIA and RIA will not be widely different. This is true if the same method is used for separation of antibody bound and unbound hormone. Differences can be seen if the methods of separation are different (Table 3). It must be empha-

TABLE 3. INFLUENCE OF THE METHOD FOR SEPARATION OF BOUND AND FREE HORMONE ON THE CROSS-REACTIVITIES OF STEROIDS IN RIA AND EIA

Hormone	Separation method for		Cross-reacting steroid	Cross-reaction in		Ref. No.
	RIA	EIA		RIA	EIA	
T	$(NH_4)_2SO_4$	DA	DHT	42	63	29
T	DCC	PEG	DHT	35	43	23
E_3	DCC	PEG	E_3G	87	60	30
			E_3SO_4	53	28	
			16 epiE_3	60	96	
			E_2	10	42	
T	DCC	Cellulose	DHT	40	23	21
P	DA	DA	5 pregnane 3,20 dione	74	75	31
			17 OHP	12	13	
F	DA	DA	Cross-reactivities in the two systems are similar			32
P	DA	DA				17
T	DCC	DA	DHT	16	12	28
T	DCC	ELISA Plate	DHT	16	25	33

DA : double antibody
DCC : Dextran coated charcoal
PEG : polyethylene glycol
T : testosterone
DHT : dihydrotestosterone
E_3 : estriol
P : progesterone
F : cortisol
E_3G : estriol 16α glucuronide
E_3SO_4 : estriol sulphate
17OHP : 17α hydroxyprogesterone
E_2 : 17 β-estradiol

sized here that, whereas a relatively non-specific antiserum can be used in RIA since the cross-reacting steroids can be removed by chromatography, EIA requires a specific antiserum since losses incurred in purification steps cannot be monitored. However, a method of immunoadsorption and extraction to get rid of cross-reacting steroids in the EIA of synthetic steroid EE_2 (ethynyl oestradiol) has been described (19).

G. Separation methods for antibody bound and unbound antigen for EIA

Because of the size and proteinaceous nature of the label, the separation methods used in steroid RIA, such as dextran-coated charcoal and salt precipitation, are not suitable for EIA. Only three methods have so far been used, i.e. double antibody, immobilised 1st or 2nd antibody (the EIA in this case is termed enzyme-linked immunosorbent assay, i.e. ELISA) or polyethylene glycol. In this laboratory, the first antibody is immobilised on microtitre ELISA plates, thus laborious and time-consuming centrifugation steps are avoided.

It is possible that, because of the use of solid phase antibody, the EIA may require higher concentrations of antibody than the corresponding RIA, but this may not always be true (Table 4).

H. Factors determining the sensitivity and precision of EIA

The characteristics of the basic ingredients, i.e. the antibody and the steroid-enzyme label, the separation system for bound and unbound hormone, the method for enzyme assay and the instrument used for assay, contribute to the sensitivity and precision of the assay.

As in the case of radioimmunoassay, a high affinity antibody, cross-reacting minimally with other steroids, will be most suitable for EIA.

The double antibody method for separation of bound and free requires repeated centrifugations and washings, and these

TABLE 4. OPTIMUM DILUTION OF ANTIBODY REQUIRED FOR RIA AND EIA OF STEROIDS

STEROID	EIA Separation method	EIA Antibody dilutions	RIA Separation method	RIA Antibody dilutions	Ref.
Testosterone	DA	1:500	$(NH_4)_2SO_4$	1:50,000	29
Estrogens	PEG	1:10600	PEG	1:24,000	30
Testosterone	PEG	1:4000	PEG	1:12,000	23
Testosterone	DA	1:15000	DCC	1:15,000	28
Cortisol	DA	1:10000	DCC	1:700	27
β-methasone	DA	1:50000	DCC	1:40,000	14
Testosterone	Microtitre plate	1:1500 (0.2 ml)	DCC	1:3000 (0.1 ml)	Author's laboratory
Estrone* glucuronide (E_1G)	-"-	1:8000 (0.2 ml)	DCC	1:12,000 (0.1 ml)	-"-
Pregnanediol* glucuronide (PdG)	-"-	1:2000 (0.2 ml)	DCC	1:4000 (0.1 ml)	-"-

*Using an ELISA reader rather than a spectrophotometer for the measurement of enzyme activity, the required dilution of E_1G antiserum and PdG antiserum are shown to be 1:30,000 and 1:15,000 respectively. The reaction volumes for the spectrophotometric measurement and ELISA reader measurement are 3 ml and 0.25 ml respectively.

DA : Double antibody PEG : Polyethylene glycol

DCC : Dextran coated charcoal

processes may lead to poor precision.

Experience from other workers have indicated that, in the case of micro-titre ELISA plates, all the wells may not have uniform adsorption capacity, thus leading to erroneous results (20). In this laboratory, the microtitration ELISA plates have been used for quantitative ELISA over a number of years and have not given rise to any problems of imprecision.

The method for enzyme assay can be modified to increase the sensitivity of EIA. Thus, using a fluorogenic substrate, i.e. p-hydroxyphenyl acetic acid for the estimation of HRP activity, instead of the conventional substrate, i.e. O-phenylenediamine, it was possible to increase the sensitivity of testosterone assay from 4 pg/tube to 0.5 pg/tube (21). The sensitivity of a cortisol EIA could also be increased four-fold using fluorimetric end-point, and twenty-fold using chemiluminescence end-point, rather than a spectrophotometric end-point (22). In this laboratory, bound enzyme penicillinase bound to ELISA plate is estimated by converting penicillin to penicilloic acid, transferring it to tubes with starch iodine and reading the absorbance after 15 minutes in a spectrophotometer. With the availability of a microtitre plate ELISA reader, (Multiskan, Flow Laboratories, U.K.) the enzyme is estimated by adding a mixture of starch-iodine-penicillin to the wells and reading the absorbance at a defined time, directly from the plate. This changeover has enabled the author to increase the sensitivity of ELISA several-fold (Fig. 1) and also to economise on the use of the antibody and label (Table 3).

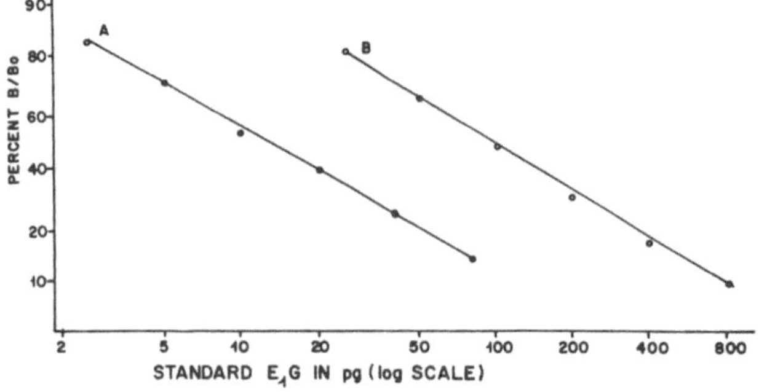

FIG.—4 Comparison of standard curves of ELISA for estrone-3 B-glucuronide standardised using A - ELISA reader and B - spectrophotometer. Intercept at 80% for A is 3.25 and for B is 26 pg. showing that assay is 8 times more sensitive than B. For A antibody dilution is 1:30,000 and label dilution is 1:2000 whereas that for B is 1:8000 and 1:700 respectively.

In the estimation of progesterone (12), a test tube containing microcrystalline cellulose or a microtitre ELISA plate were used for immobilising antibody. The plate assay was found to be four times more sensitive than the tube assay.

The precision of the assay will also depend upon the range of optical density readings noted in the assay. Table 5 gives details of readings observed in some of the assays.

TABLE 5. RANGE OF STEROID LEVELS THAT CAN BE MEASURED AND CORRESPONDING RANGE OF OPTICAL DENSITIES BY SOME EIA SYSTEMS

Steroid	Range of Conc./tube or well	Range of O.D. readings	Ref.
P	125 - 4000 pg	0.1 - 0.44	24
T	60 - 1500 pg	0.05- 0.5	23
F	1 - 80 ng	0.2 - 0.8	25
P	20 - 5000 pg	0.15- 0.45 (tube assay)	12
	5 - 5000 pg	0.03- 0.13 (microplate assay)	
P	25 - 5000 pg	0.22- 0.42	17
T	12.5 - 400 pg	0.3 - 1.4 on spectro-photometer	Author's laboratory
E_1G	25 - 800 pg		
PdG	250 - 16000 pg		
E_1G	2.5 - 80 pg	0.3 - 1.4 on ELISA reader	-"-
PdG	125 - 4000 pg		
P	25 - 1600 pg		

P : Progesterone
T : Testosterone
F : Cortisol
E_1G : Estrone-3α-glucuronide
PdG : Pregnanediol-3α-glucuronide

When the difference between minimum and maximum readings is too small, a spectrophotometer with high resolution, capable of recording small differences in readings accurately, will be required for a precise assay.

III. Fluoroimmunoassays (FIA)

A. Types of FIAs

Although labelling of a ligand or an antibody with a fluorescent molecule to localise biologically active substances in tissues has been in practice for decades, its use in immunoassays for steroid hormones in biological fluids has only recently been exploited. Two types of assay have been published, those requiring separation of antibody bound hormone from unbound hormone and the others not requiring such a step. The latter type of assays, called polarization fluoroimmunoassays (PFIA), make use of the observation that fluorescence polarisation arises when a fluorochrome is irradiated by polarised light, it absorbs energy and returns to its nonexcited state by emitting a light photon. The photon is also polarised because it is re-emitted according to the orientation of the emitting molecule. Only small freely tumbling fluorochrome molecules in solution exhibit fluorescence polarization. The property is also exhibited by fluorochrome labelled haptens. However, when the labelled haptens combine with the antibody, the molecule becomes large, rate of tumbling goes down and fluorescence polarization is modified accordingly.

Another type of fluoroimmunoassay, i.e. fluorescence quenching immunoassay (FQI) uses the fact that the fluorescein dye has marked quenching in fluorescence when bound to an antibody. This type of assay also does not require separation of antibody bound hormone from unbound hormone. FQI has been developed for the estimation of cortisol (34).

B. Labels used in fluoroimmunoassays

Fluorochromes used in methodologies published so far are fluorescein, fluorescein amine, fluorescein thiocarbamyl ethylenediamine, phenanthroimidazole-2 amine and 4 methyl-umbelliferyl 3 acetic acid. There are no known distinct advantages of one fluorochrome over the other.

The fluorochromes are conjugated to the steroid derivative through mixed anhydride reaction or through carbodiimide and N-hydroxysuccinimide, as described for conjugating enzymes to steroids. Virtually no data exists on the stability of the labels but it is assumed that these labels should have a long shelf life. Thus, fluorescein labelled cortisol could be stored in good condition for at least 4 months when preserved at $4^{o}C$ or at $-20^{o}C$ (35).

C. Sensitivity of fluoroimmunoassays

The sensitivity of the fluoroimmunoassays is generally low as compared to RIA and EIA and only nanomolar quantities can be measured with accuracy and precision. Because of this, assays have only been published for steroids which are present in the circulation in relatively high concentrations, i.e. cortisol (34-38), deoxycortisol (39), and progesterone (40, 41). A method of attaching fluorochrome to steroid has been introduced (42), whereby, in a fluoroimmunoassay for 5α-dihydrotestosterone (5α-DHT), polylysine labelled with 5 methyl umbelliferyl-3 acetic acid (4 MU-3 acetic acid) was used and the sole terminal carboxyl group of polylysine was conjugated to 3 amine derivative of 5α-DHT. Thus, per molecule of polylysine and, in turn, per molecule of 5α-DHT, 35 molecules of 4 MU-3 acetic acid could be attached; a sensitivity comparable to that of RIA was achieved.

D. Problems associated with fluoroimmunoassays

One inherent drawback in the fluoroimmunoassay is that serum itself has endogenous fluorescence which gives rise to a high blank, thus leading to a low 'signal to noise ratio'. This problem has been minimised by extracting the sample with a solvent which does not extract fluorogenic substances (38-40, 42) or by using a solid phase antibody (37). When a non-extraction method was developed, binding to serum protein was prevented by adding sodium salicylate (35, 37) or sodium dodecyl sulphate (36), the latter also helping to clear turbid samples.

The polarization fluoroimmunoassays, no doubt attractive because they do not require a separation step, may still not be very popular since they require precise instrumentation, i.e. a polarization spectrofluorimeter with attached oscilloscope for integrating the signals over a period of time. Such an instrument does not normally form a part of routine equipment in a diagnostic laboratory.

The sensitivity of currently available FIA is not suitable for steroids present in the circulation in picomolar or femtomolar concentrations. However, development of assays using fluorescently labelled polylysine conjugates of haptens (42) will lead to development of sensitive assays.

In general, fluoroimmunoassays require antibodies in concentrations much higher than those needed for conducting RIA and EIA, and thus may not be economical.

IV. Chemilumiscent Immunoassays (CIA)

Some aminophthalhydrazides participate in simple oxidation reactions to produce light with high quantum efficiencies. The reaction is called chemiluminescence and the molecules exhibiting this property are called chemiluminescent molecules. Possibility of their use as labelling material for ligands in competitive protein binding methods has been suggested (43) and their use in the estimation of thyroxine was also described (44). Since then, chemiluminescent immunoassays (CIA) have been described for a number of haptens including steroids. This method has been discussed in detail elsewhere in this volume, on page 105.

Chemiluminescent labels used in steroid CIAs

Although the chemiluminescence property of luminol (5 amino-2, 3-dihydrophthalazine,1-4 dione) has been known for over 50 years, it is its isomer isoluminol (6 amino 2,3-dihydrophthalazine 1-4 dione) or its derivatives that are used as a label in CIA. Ligands of interest are attached through an alkyl-bridging group to the amino residue of isoluminol. Alkylation of amino residue increases the quantum efficiency of isoluminol to the same value as luminol. A similar substitution on luminol decreases its quantum efficiency substantially.

The oxidation reaction of isoluminol is achieved mainly by nascent oxygen provided by the reaction of microperoxidase on hydrogen peroxide.

Several derivatives of isoluminol have been synthesised, i.e. aminoethyl isoluminol (AEI), aminoethyl-ethyl isoluminol (AEEI), amino butyl isoluminol (ABI), aminobutyl ethyl isoluminol (ABEI), aminopentyl ethyl isoluminol (APEI) and amino hexyl ethyl isoluminol (AHEI). Although all of them are found to be suitable labels (45), ABEI seems to be the popular choice.

The labelling of steroid ligands with chemiluminescent molecules is achieved through the same procedure as is used for steroid-enzyme or steroid-fluorochrome conjugates. These are the mixed anhydride reaction and the carbodiimide reaction. Purification is achieved by thin-layer chromatography.

Types of CIAs

Two types of assay have been developed so far, the heterogeneous assay requiring separation of bound and free hormone and the homogeneous assay, not requiring this step. In most of the published assays on CIA, the heterogeneous assays employ a solid phase first antibody, using polystyrene test tubes (46-48) or immunobeads (49) as a solid phase. The observation that the light intensity of a hapten-chemiluminescent conjugate is enhanced on its combination with the hapten antibody (44) opened up the possibility of homogeneous CIA for haptens, including steroids. Such assays have been developed for the estimation of total estrogens (50) and estriol-16α-glucuronide (51) in urine.

The sensitivity of CIA is comparable to that of RIA but the assays are less precise than RIA, the interassay coefficient of variation in CIA being 15%-20% as against 10%-15% for RIA.

Recently, CIAs have been developed for progesterone (46, 52), 17β-estradiol (49), and cortisol (53) in serum/plasma and estriol 16α-glucuronide (51), pregnanediol glucuronide (47), testosterone glucuronide (48), and total estrogens (50), in urine. In view of the high sensitivity of CIA, the stability of label for relatively longer periods and the short time taken for the performance of the assays, CIAs are likely to occupy a prominent position among the diagnostic tools.

V. Applications of Non Isotopic Immunoassays (NIIA) of Steroids in Endocrinology

It is about a decade now since the NIIA were developed as an alternative method to RIA of steroid hormones. Perusing important journals in the fields of endocrinology, obstetrics and gynaecology and reproduction, where steroid assays have an important application, one notices that NIIA have not been used as a routine tool in hormone assay laboratories. There may be several reasons for this, i.e. unwillingness to change over from well established methods to relatively new methods, need for buying new equipment such as polarization fluorimeters and luminometers, need to train technicians in the new assay methods etc. However, the most important constraint is non-availability of ready-to-use kits on a large scale. Kits for protein hormones and for thyroid hormones are now available but such kits for steroids are either not available or not widely publicised, a situation which might change when the demand increases. The following paragraphs review the available information on application of NIIA for steroid hormones.

The penicillinase-linked ELISA developed in this laboratory makes use of the microtitration ELISA plates for separation of antibody bound and unbound hormones. The assays have a uniform protocol for protein and steroid hormones and are rapid to perform (3 hours for completion) (54). The sensitivity, precision, practicality and rapidity of the system has enabled these assays to be applied for management of patients on gonadotropin therapy, for rapid screening of patients on ovulation inducing agents to record their response to therapy, for monitoring the ovarian function of patients recruited for '*in vitro* fertilization' and for studying the testicular function of cryptorchid testes. Some examples are given below.

Figure 2 shows a case of treatment with gonadotropins. Decision to determine the dose of Pergonal$^{(R)}$, discontinuation of Pergonal and administration of hCG was taken on the basis of urinary estrone glucuronide profile prepared by ELISA, cervical mucus monitoring and ultrasound scanning of ovaries. This woman became pregnant on therapy.

Figure 3 shows the response of a subject who developed hyperstimulation when on gonadotropin therapy. In this case ultrasound measurement of ovarian follicle size was not possible since the patient was 'unable to fill the bladder'. Excessive response was suspected on the basis of E_1G profile and hCG injection in this case was withheld, taking into consideration the high values of urinary E_1G and the size of the ovary by palpation.

In the Infertility Clinic of the Institute for Research in Reproduction, the possibility of corpus luteum inadequacy in selected cases of infertility has been proven or ruled out on the basis of pregnanediol glucuronide (PdG) excretion in urine, assayed by ELISA. Figure 4 shows the PdG excretion in 3 cases of suspected corpus luteum inadequacy.

An '*in vitro*' fertilization programme was started in 1982, in collaboration with a leading teaching hospital (KEM Hospital)

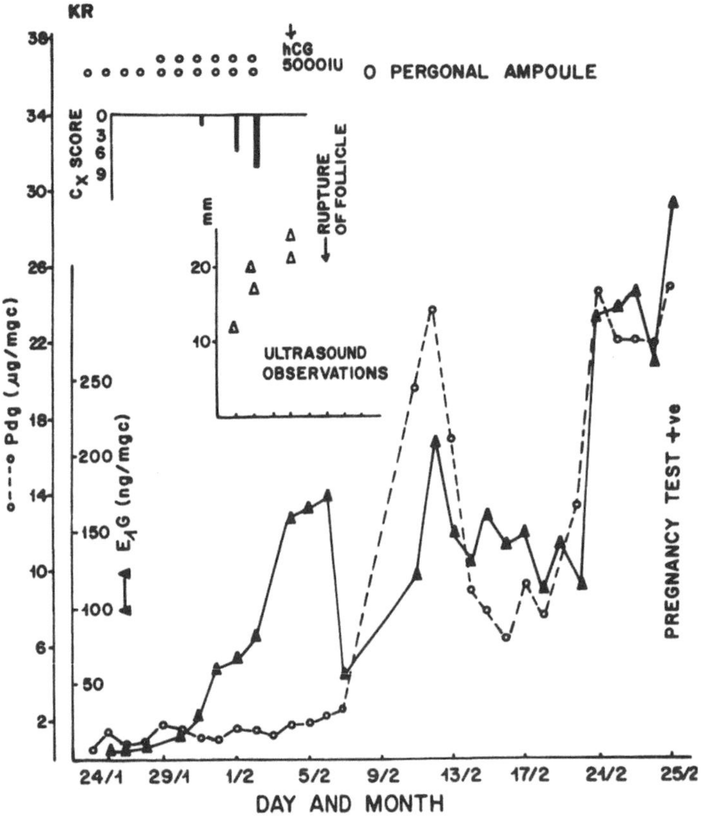

FIG.-2 KR, a case of secondary amenorrhoea was started on 1 ampoule of Pergonal. In view of no rise in E_1G by day 4 of treatment, the dose was increased to 2 ampoules per day. Pergonal was stopped when response in terms of increase in follicular size (16 mm & 20 mm), cervical mucus score (8) and E_1G (80 ng/mg c) was adequate. hCG (5000 i.u.) was given two days later. 15 days after the rupture of follicle, E_1G and PdG were still high and beyond the range of peak luteal phase values, indicating the possibility of pregnancy, which was confirmed 4 days later by pregnancy test.

in Bombay. The time of laparoscopic aspiration of the oocytes was decided taking into consideration the hormonal profile (estrone-3α-glucuronide and LH in urine by ELISA) cervical

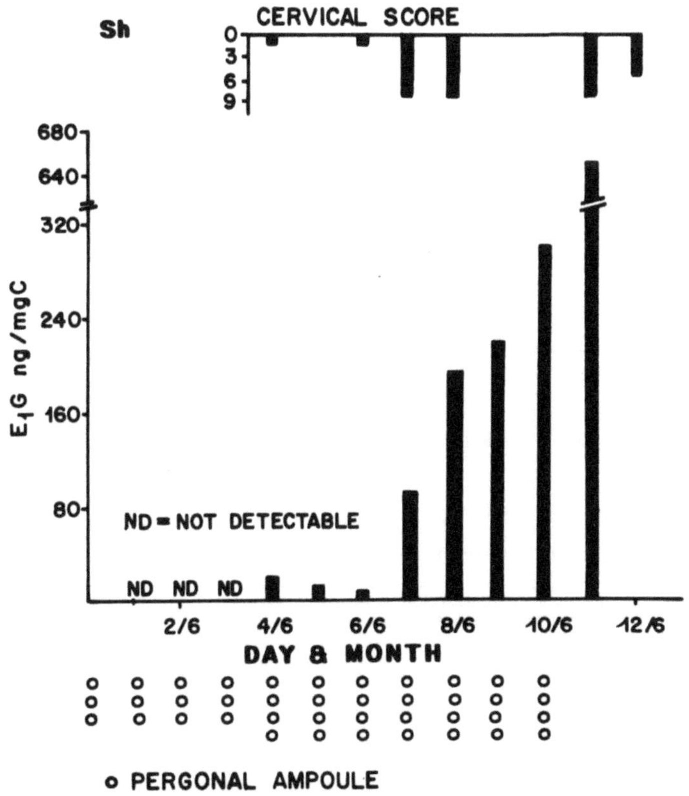

FIG.-3 Sh, a case of secondary amenorrhoea, was known to be a poor responder to gonadotropin therapy on previous two occassions. Urine samples from 31/5 to 11/6 were sent for analysis of urinary E_1G estimations before hCG injection was planned. High levels of E_1G (652 ng/mg creatinine) was indicative of hyperstimulation, hence hCG was not given.

mucus score and ultrasound scanning of ovary. The rate of harvesting mature oocytes in 1983, when the hormone assay results were available, retrospectively, and in 1984, when hormone assay results were available prospectively, in addition to clinical parameters, are given in Table 6. Results show that the availa-

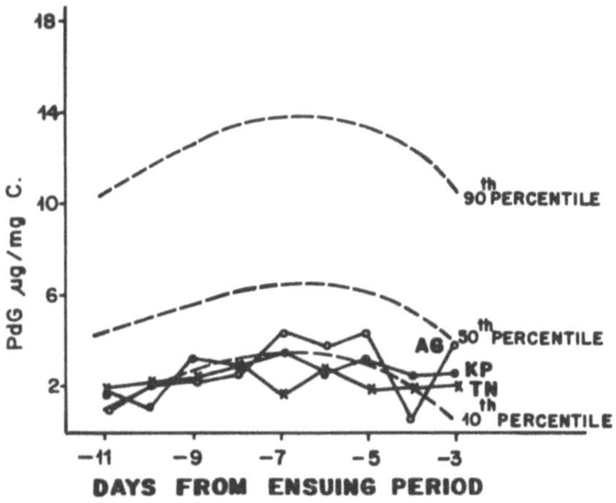

FIG.-4 Excretion of pregnanediol 3 glucuronide (PdG) in the urines of 3 infertile women (AG, TN, KP) with suspected corpus luteum inadequacy. Broken lines indicate 90th, 50th & 10th percentile of values derived from data on 45 regularly menstruating women.

bility of the hormone assay results on time have increased the chances of collecting mature oocytes for in vitro fertilization.

In the Genetic Clinic of the Institute, the availability of ELISA for serum testosterone has helped in the management of cases of cryptorchid testes. A good response to hCG in boys below 3 years of age indicates the possibility of bringing the testes into the scrotum with hCG therapy, a poor response indicates the need for surgery. The hCG stimulation test in older boys after surgery for cryptorchid testes helps in the counselling of parents (Table 7).

The EIA for testosterone with the fluorometric end point developed (21), is sensitive enough to detect salivary testosterone in the female and has been used to assess the androgenic status of some infertile women. Thus, occurrence of androgen excess in a hirsute oligomenorrhic woman could be established with the use of this assay. Notably, the plasma testosterone concentration in this case was within normal limits. The EIA

TABLE 6. RESULTS OF LAPAROSCOPIC ASPIRATION AND IN VITRO FERTILIZATION (IVF) IN CASES RECRUITED FOR IVF PROGRAMME

	Cases recruited in		
	1983		1984
	Therapy		
	Fertyl	Pergonal	FSH
No. of aspirations	5	9	13
(No. of cases)	(6)	(6)	(12)
No. of follicles aspirated	14	12	30
No. of Oocytes recovered			
pre ovulatory	0	3	13
immature	0	4	8
atretic	4	0	1
No. of oocytes matured	0	3	13
No. of eggs fertilized	0	0	9
No. of eggs cleaved	0	0	2
Embryo transfer at 8 cell stage	0	0	2

In 1983 decision for discontinuing Pergonal, administering hCG and laparoscopic aspiration was based on ultrasound scanning of ovary and cervical mucus score. In 1984 additional parameter for follicular development (estrone-3-glucuronide in urine estimated by ELISA) was introduced.

has also been developed for synthetic steroids norethisterone and ethynyl estradiol (12) and these assays have been applied for the study of pharmacokinetics of these steroids in women

TABLE 7. USE OF SERUM TESTOSTERONE ESTIMATION IN THE MANAGEMENT OF BOYS WITH CRYPTORCHID TESTES

Case no.	Age	Serum testosterone ng/ml		Remarks
		Pre hCG	Post hCG	
1	1 yr. 6 mo.	1.2	1.7	Surgical management
2	3 yr.	1.1	1.3	"
3	1 yr. 3 mo.	1.3	2.7	hCG therapy
4	15 yr.	9.5	-	Good prognosis

on steroidal contraceptives. According to the report (12), these cost-effective assays 'could well be particularly useful for processing the samples generated by field studies of new contraceptive formulations in developing countries'.

The solid phase chemiluminescence immunoassays for plasma estradiol-17β have been successfully employed (49) to monitor gonadotropin therapy for induction of ovulation. The values for E_2-17β determined by CIA in these cases agreed well with those obtained by RIA using either an iodine label or a tritium label.

Enzyme immunoassay of progesterone in milk has been successfully applied (10) in the diagnosis of pregnancy in cows. Although the precision of this direct assay suitable for field studies is not so reliable (C.V. between 22%-28%), the assay is shown to be 95%-98% accurate in diagnosing non-pregnancy. For pregnancy, the accuracy improves if 4 to 7 consecutive samples, between day 20 to day 24 after insemination, are analysed.

Future of Non-Isotopic Immunoassays for Steroid Hormones

There is no doubt that non-isotopic techniques for steroid immunoassays will be the method of choice by the routine hormone assay laboratories in the near future. Development of monoclonal antibodies to overcome the problem of cross-reactivities, and of direct non-extraction type of methods will greatly simplify these sensitive and specific assays. The ultimate goal will of course be to develop homogeneous assays not requiring bound and free hormone separation, thus paving the way to complete automation of the methods. Moreover, it is likely that semiquantitative methods, which could be carried out in a clinician's practice, may also be developed within the next few years.

References

1. Van Weeman, B.K., Schuurs, A.H.W.M.: Immunoassay using antigen-enzyme conjugates. FEBS Lett. 15, 232-236 (1971).

2. Engvali, E., Perlmann, P.: Enzyme-linked immunosorbent assay (ELISA). Quantitative assay of immunoglobulin G. Immunochemistry 8, 871-874 (1971).

3. Erlanger, B.T., Borek, F., Beiser, S.M., Lieberman, S.: Steroid-protein conjugates. J. biol. Chem. 228, 713-727 (1957).

4. Goodfriend, T.L., Levine, L., Fosman, G.D.: Antibodies to bradykinn and angiotensin: a use of carbodiimides in immunology. Science 144, 1344-1346 (1964).

5. Anderson, G.W., Zimmerman, J.E., Callahan, F.M.: The use of N-hydroxy succinimide in peptide synthesis. J. Am. chem. Soc. 86, 1839-1842 (1964).

6. Exley, D., Abuknesha, R.: A highly sensitive and specific enzyme immunoassay method for oestradiol-17β. FEBS Lett. 91, 162-165 (1978).

7. Rajkowski, E.M., Cittanova, N., Desfosses, B., Jayle, M.F.: The conjugation of testosterone with horse radish peroxidase and a sensitive enzyme immunoassay for the conjugate. Steroids 29, 701-713 (1977).

8. Comoglio, S., Celada, F.: An immunoenzymatic assay of cortisol using E. Coli β-galactosidase as label. J. Immunol. Methods 10, 161-170 (1976).

9. Rajkowski, K.M., Cittanova, N.: The efficiency of different coupling procedures for the linkage of estriol-16α-glucuronide, estrone 3α-glucuronide and pregnanediol 3α-glucuronide to four different enzymes. J. Steroid Biochem. 14, 861-866 (1981).

10. Chang, C.F., Estergreen, V.L.: Development of a direct enzyme immunoassay of milk progesterone and its application to pregnancy diagnosis in cows. Steroids 41, 171-195 (1983).

11. Dray, F., Andrien, J.M., Renond, F.: A sensitive immunoassay of progesterone using β-galactosidase as label. In: "Immunoenzymatic Techniques", Eds.Feldmann, G., Druet, P., Bignon, J., Avrameas, S., North Holland Publishing Co., Amsterdam, pp. 183-189 (1975).

12. Sauer, M.J., Foutkes, J.A., Cookson, A.D.: Direct enzyme immunoassay of progesterone in bovine milk. Steroids 38, 45-54 (1981).

13. Joyce, B.G., Read, G.F., Riad-Fahmy, D.: Enzyme immunoassay for progesterone and oestradiol: a study of factors influencing sensitivity. In: "Radioimmunoassay and Related Procedures in Medicine", International Atomic Energy Authority, Vienna, pp. 289-295 (1978).

14. Kominami, G., Yamauchi, A., Ishihara, S., Kono, M.: An enzyme immunoassay for plasma betamethasone. Steroids 37, 303-314 (1981).

15. Arakawa, H., Maeda, M., Tsuji, A., Kambegawa, A.: Chemiluminescence enzyme immunoassay of dehydroepiandrosterone and its sulphate using peroxidase as label. Steroids 38, 453-464 (1981).

16. Bosch, A.M.G., Dikhuizen, D.M., Schuurs, A.H.W.M., Van Weeman, B.K.: Enzyme immunoassay for total estrogens in pregnancy plasma or serum. Clin. chim. Acta 89, 59-70 (1978).

17. Nakao, T.: Practical procedure for enzyme immunoassay of progesterone in bovine serum. Acta endocr. 93, 223-227 (1980).

18. Ogihara, T., Miyai, K., Nishi, K., Kaichiro, I., Kumahara, Y.: Enzyme-labelled immunoassay for plasma cortisol. J. clin. Endocr. Metab. 44, 91-95 (1977).

19. Turkes, A., Dyas, J., Read, G.F., Riad-Fahmy, D.: Enzyme-immunoassay for specific determination of the synthetic estrogen ethynyl oestradiol in plasma. Clin. Chem. 27, 901-905 (1981).

20. Voller, A.: Heterogeneous enzyme immunoassay and their applications. In: "Enzyme-immunoassay", Ed. Maggio, E.T., CRC Press, Florida, USA, pp. 181-196 (1980).

21. Turkes, A.O., Turkes, A., Joyce, B.G., Riad-Fahmy, D.: A sensitive enzyme immunoassay with a fluorimetric end point for the determination of testosterone in female plasma and saliva. Steroids 35, 89-107 (1980).

22. Arakawa, H., Maeda, M., Tsuji, A.: Chemiluminescence enzyme immunoassay of cortisol using peroxidase as label. Analyt. Biochem. 97, 248 (1979).

23. Osterman, T.M., Juntunen, K.O., Gothori, G.D.: Enzyme-immunoassay of testosterone in plasma with use of polyethylene glycol to separate antibody-bound and free hormone. Steroids 34, 575-579 (1979).

24. Joyce, B.G., Read, G.F., Riad-Fahmy, D.R.: A specific enzyme immunoassay for progesterone in human plasma. Steroids 29, 761-770 (1977).

25. Kobayashi, Y., Ogihara, T., Amitani, K., Watanabe, F., Kiguchi, T., Ninomiya, I., Kumahara, Y.: Enzyme immunoassay for cortisol in serum using cortisol-21 amine. Steroids 32, 137-144 (1978).

26. Turkes, A., Turkes, A.O., Joyce, B.G., Read, G.F., Riad-Fahmy, D.R.: A sensitive solid phase enzyme immunoassay for testosterone in plasma and saliva. Steroids 33, 347-359 (1979).

27. Kominami, G., Fujisaka, I., Yamaguchi, A., Kono, M.: A sensitive enzyme immunoassay for plasma cortisol. Clin. chim. Acta 103, 381-391 (1980).

28. Joshi, U.M., Shah, H.P., Sudhama, S.P.: A sensitive and specific enzyme immunoassay for serum testosterone. Steroids 34, 35-46 (1979).

29. Tateishi, K., Yamamota, H., Ogihara, T., Hayashi, C.: Enzyme immunoassay of serum testosterone. Steroids 30, 25-32 (1977).

30. Osterman, T.M., Juntunen, K.O., Gothoni, G.D.: Enzyme immunoassay of estrogen-like substances in plasma with polyethylene glycol as precipitant. Clin. Chem. 25, 716-718 (1979).

31. Joyce, B.G., Wilson, D.W., Read, G.F., Riad-Fahmy, D.R.: An improved enzyme immunoassay for progesterone in human plasma. Clin. Chem. 24, 2099-2102 (1978).

32. Nakao, T., Tamamura, F., Tsunoda, N., Kawata, K.: Double antibody enzyme immunoassay of cortisol in bovine plasma. Steroids 38, 111-120 (1981).

33. Joshi, U.M., Shah, H.P., Sankolli, G.M.: Penicillinase as a marker in enzyme-linked immunosorbent assays for steroid hormones. J. Steroid Biochem. 19, 419-421 (1983).

34. Kobayashi, Y., Tsubota, N., Miyai, K., Watanabe, F.: Fluorescence quenching immunoassay of serum cortisol. Steroids 36, 177-183 (1980).

35. Pourfarzaneh, M., White, U.M., Landon, J., Smith, D.S.: Cortisol directly determined in serum by fluoro immunoassay with magnetizable solid phase. Clin. Chem. 26, 730-733 (1980).

36. Kobayashi, Y., Tsubota, N., Miyai, K., Watanabe, F.: Direct fluorescence polarization immunoassay of serum cortisol. Steroids 34, 829-834 (1979).

37. Kobayashi, Y., Yahata, M., Watanabe, F., Miyai, K.: A solid phase fluoroimmunoassay of serum cortisol. J. Steroid Biochem. 16, 521-524 (1982).

38. Al-Ansari, A.A., Smith, D.S., Landon, J.: Polarization fluoroimmunoassay of serum and salivary cortisol. J. Steroid Biochem. 19, 1475-1480 (1983).

39. Al-Ansari, A.A., Massoud, M., Perru, L.A., Smith, D.S.: Polarization fluoroimmunoassay of 11-deoxy-cortisol in serum and saliva. Clin. Chem. 29, 1803-1805 (1983).

40. Bacigalupo, M.A., Ius, A., Meroni, G., Saita, R., Parini, C.: Fluoroimmunoassay of progesterone in plasma by phenanthroimidazole-2 amine labelling. J. Steroid Biochem. 19, 1661-1664 (1983).

41. Allman, B.L., Short, F., James, V.H.: Fluoroimmunoassay of progesterone in human serum or plasma. Clin. Chem. 27, 1176-1179 (1981).

42. Exley, D., Ekeke, G.I.: Fluoroimmunoassay of 5α-dihydrotestosterone. J. Steroid Biochem. 14, 1297-1302 (1981).

43. Schroeder, H.R., Vogelhut, P.O., Carrico, R.J., Boguslaski, R.C., Buckler, R.T.: Competitive protein binding assay for biotin monitored by chemiluminescence. Anal. Chem. 48, 1933-1942 (1976).

44. Schroeder, H.R., Boguslaski, R.C., Carrico, R.J., Buckler, R.T.: Monitoring specific protein binding reactions with chemiluminescence. Methods in Enzymol. 57, 424-445 (1978).

45. Pazzagli, M., Kim, J.B., Messeri, G., Martinazzo, G., Kohen, F., Franceschetti, F., Moneti, G., Salerno, R., Tommasi, A., Serio, M.: Evaluation of different progesterone-isoluminol conjugates for chemiluminescence immunoassay. Clin. chim. Acta 115, 277-286 (1981).

46. Kohen, F., Kim, J.B., Lindner, H.R., Collins, W.P.: Development of a solid-phase chemiluminescence immunoassay for plasma progesterone. Steroids 38, 73-88 (1981).

47. Eshhar, Z., Kim, J.B., Barnard, G., Collins, W.P., Gilad, S., Lindner, H.R., Kohen, F.: Use of monoclonal antibodies to pregnanediol-3-glucuronide for the development of a solid phase chemiluminescence immunoassay. Steroids 38, 89-110 (1981).

48. Vannuckhi, P.L., Messeri, G., Bolelli, G.F., Pazzalli, M., Masala, A., Serio, M.: A solid phase chemiluminescent immunoassay for testosterone glucuronide in diluted urines. J. Steroid Biochem. 18, 625-629 (1983).

49. De Boever, J., Kohen, F., Vandekerckhove, D.: Solid phase chemiluminescence immunoassay for plasma E_2 - 17β during gonadotropin therapy compared with two radioimmunoassays. Clin. Chem. 29, 2068-2071 (1983).

50. Messeri, G., Caldini, A.L., Boleli, G.F., Pazzagli, M., Tommasi, A., Vannuchi, P.L., Serio, M.: Homogeneous luminescence immunoassay of total estrogens in urine. Clin. Chem. 30, 653-657 (1984).

51. Kohen, F., Kim, J.B., Barnard, G., Lindner, H.R.: An assay for urinary estriol - 16 glucuronide based on antibody enhanced chemiluminescence. Steroids 36, 405-419 (1980).

52. Pazzagli, M., Kim, J.B., Messeri, G., Martinazzo, G., Kohen, F., Franceschetti, F., Tommasi, A., Salerno, R., Serio, M.: Luminescent immunoassay for progesterone in a heterogeneous system. Clin. chim. Acta 115, 287-296 (1981).

53. Kohen, F., Pazzagli, M., Kim, J.B., Lindner, H.R.: An immunoassay for cortisol based on chemiluminescence. Steroids 36, 421-437 (1980).

54. Joshi, U.M., Shah, H.P., Sankolli, G.M., Khatkhatay, M.I.: Development and application of penicillinase linked EIA in reproductive medicine in clinical biochemistry - Principles and practice. Proceedings of the Second Asian and Pacific Congress of Clinical Biochemistry, Singapore, 1982, Eds. Eng, A.S., Garcia-Webb, P., pp. 257-262 (1983).

CONTRIBUTORS

Numbers in parentheses indicate the page on which the authors articles begin

S. Iwasa, Chemistry Laboratory, Central Research Division, Takeda Chemical Industries Ltd., Osaka 532, Japan (19).

J. H. Jackson, Department of Psychiatry M-003, School of Medicine, University of California at San Diego, La Jolla, California 92093, U.S.A. (45).

U. M. Joshi, Institute for Research in Reproduction (ICMR), Parel, Bombay 400 012, India (151).

M. F. Laker, Department of Clinical Biochemistry, Royal Victoria Infirmary, Newcastle upon Tyne, NE 1 4LP, U.K. (1).

S. Matsuura, Department of Chemistry, College of General Education, Nagoya University, Nagoya 464, Japan (91).

T. Nagatsu, Department of Biochemistry, Nagoya University School of Medicine, Nagoya 466, Japan (91).

M. Sawada, Laboratory of Cell Physiology, Department of Life Chemistry, Graduate School of Nagatsuta, Tokyo Institute of Technology, Yokohama 227, Japan (91).

T. Sugimoto, Department of Chemistry, College of General Education, Nagoya University, Nagoya 464, Japan (91).

T. R. Trinick, Department of Clinical Biochemistry, Royal Victoria Infirmary, Newcastle upon Tyne, NE1 4LP, U.K. (1).

W. G. Wood, Klinische Laboratorien, Klinik für Innere Medizin, Medizinische Hochschule Lübeck, Ratzeburger Allee 160, 2400 Lübeck 1, Federal Republic of Germany (105).

T. Yamaguchi, Laboratory of Cell Physiology, Department of Life Chemistry, Graduate School of Nagatsuta, Tokyo Institute of Technology, Yokohama 227, Japan (91).

SUMUSU IWASA was born in Japan on December 21st, 1945. In 1968 he graduated from the School of Science, Kyoto University. From 1968 to 1970 he followed a post-graduate course (Master course) at the Department of Chemistry, Kyoto University School of Science. Since 1970 he has been a Research Associate at the Chemistry Laboratory, Central Research Division, Takeda Chemical Industries Ltd., Osaka, Japan. From 1982 to 1983 he was Visiting Researcher at the Department of Internal Medicine, Utah University School of Medicine, Utah, U.S.A. Dr. Iwasa has a Major in Biochemistry and Immunology and obtained a Ph.D. from Tokyo University in 1981. He is a Member of the Japanese Biochemical Society, the Japanese Pharmaceutical Society, the Japanese Society for Endocrinology and the Japanese Society for Virology.

DR. JACKSON received his B.S. in Zoology with a Chemistry minor from the University of Rhode Island and his Ph.D. from the Department of Neuroscience at the University of California at San Diego (UCSD). His research emphasis and expertise are in the neurochemistry and neuropharmacology of the cerebral cortex, midbrain, and striatum. Other interests include aspects of the limbic system and regeneration in the CNS. Dr. Jackson has been a postdoctoral Research Associate in the Department of Psychiatry at UCSD with which he is presently reassociated, and the Department of Biochemistry at the Louisiana State University

Medical Center in New Orleans, where he was located when this manuscript was prepared. He uses isoelectric focusing to purify and study brain enzymes. His present research focuses on studying the relationship between kinase reactions (protein kinases and adenylate kinase) and phosphorylated monooxygenases (tryptophan and tyrosine hydroxylase) in the midbrain and striatum.

DR. USHA M. JOSHI graduated in Chemistry from Bombay University in 1954 and obtained a Doctorate in Biochemistry in 1970. Her main research interests are steroidal contraception, infertility and development of hormone assays that can be used in moderately equipped laboratories of the developing countries. Dr. Joshi was a member of the WHO Task Force on 'Interaction of Nutrition and Infectious Diseases with Steroidal Contraceptives' and has made important contributions in this field. She also served as a member on the WHO Steering Committee for Standardization of Immunoassays for Reproductive Hormones', and has done pioneering work in the field of ELISA of reproductive hormones and this work has been highly acclaimed at several National and International meetings in which she participated. She was awarded the prestigeous Indian 'VASVIK' award for women scientists for the successful transfer of the technical know-how to the industrial sector for the development of an ELISA kit for pregnancy detection. Dr. Joshi is a recognised guide for Ph.D. programmes at the Bombay University and has published over sixty scientific publications in many journals.

After qualifying in Medicine and Biochemistry in 1969, DR. LAKER undertook training in Pathology, specialising in Clinical Biochemistry, at St. Thomas's Hospital Medical School. He gained the M.R.C.Path. in 1976 and M.D. in 1979. After a year as Medical Research Council Travelling Fellow in Gastroenterology at the University of California, San Diego, he took up his present post of Senior Lecturer in Clinical Biochemistry and Metabolic Medicine , University of Newcastle upon Tyne, and Consultant in Clinical Biochemistry, Newcastle Health Authority in 1980. His research interests include oxalate metabolism, intestinal absorption and the application of analytical techniques to Clinical Medicine.

SADAO MATSUURA graduated from Nagoya University with a B.Sci. degree in 1950 and from the Australian National University with a Ph.D. degree in 1961. He is Professor of Chemistry at Nagoya University.

TOSHIHARU NAGATSU obtained his M.D. and Doc. Med. Sci. degrees at Nagoya University. From 1966 to 1976 he was Professor of Biochemistry at Aich-Gakuin University of Dentistry, and since 1976 he has been Professor of Cell Physiology in the Department of Life Chemistry at the Graduate School of Nagatsuta in the Tokyo Institute of Technology. Since 1984 he has been Professor of Biochemistry at Nagoya University School of Medicine. He was NIH fellow from 1962 to 1964, Visiting Professor at the

University of Southern California from 1967 to 1968, and Visiting Scientist at Roche Institute of Molecular Biology from 1972 to 1973.

MAKOTO SAWADA graduated from the Department of Chemistry, Faculty of Science, the Tokyo Institute of Technology with a B.Sci. degree in 1981, from the Department of Life Chemistry, Graduate School at Nagatsuta, the Tokyo Institute of Technology with an M.Sci. degree in 1982, and is now working at the Ph.D. course of the Department of Life Chemistry of the Graduate School at Nagatsuta in the Tokyo Institute of Technology.

TAKASHI SUGIMOTO graduated from Nagoya University with a B.Sci. degree in 1965, from the Australian National University with a Ph.D. degree in 1971. He is an Associate Professor of Chemistry at Nagoya University.

After obtaining an honours B.Sc. in Pathology in 1975, DR. TRINICK graduated in Medicine from The Queens University of Belfast in 1977. He spent a year in Histopathology followed by three years in general medicine during which time he obtained the M.R.C.P. (UK). At present he is Senior Registrar in Clinical Biochemistry and Metabolic Medicine at the Royal Victoria Infirmary, Newcastle upon Tyne, seconded from Northern Ireland. Research interests include human nutrition and glucose metabolism.

DR. WOOD was born on 23.04.1944 in Colne, Lancashire, England, and obtained a B.Sc. (Hons.) in Biochemistry in 1966, and a Ph.D. in Anatomy in 1972, from Leeds University. He is a Member of the Institute of Biology since 1972, has an Anerkennung als Klinischer Chemiker, Munich, F.R. Germany since 1979, and received a Habilitation in Clinical Chemistry from the Medizinische Hochschule, Lübeck in 1984. From 1972 to 1975 Dr. Wood worked in the Department of Chemical Pathology and Experimental Surgery at St. Bartholomew's Hospital, London, after which he moved to Munich where he worked in the Institute of Clinical Chemistry and Clinical Biochemistry at the Technical University (1975-1976) and in the Department of Clinical Chemistry and Endocrinology at the Ludwig-Maximilian University (1976-1980). Since 1980, Dr. Wood has been in charge of the diagnostic and research laboratories of the Clinic for Internal Medicine in the Medical University of Lübeck. The present emphasis lies in the development and implementation of non-isotopic immunoassays, especially luminescence immunoassays, with which, at the time of writing, over 40 analytes can be measured.

TOKIO YAMAGUCHI graduated from the International Christian University in 1977 with a B.Sci. degree, from the Department of Life Chemistry of the Graduate School of Nagatsuta, the Tokyo Institute of Technology with an M.Sci. degree in 1979, and with a D.Sci. degree in 1982. He has been on the research staff in the Department of Biochemistry at the Central Research Laboratories of the Yamanouchi Pharmaceutical Company.

SUBJECT INDEX

Bei Fragen zur Produktsicherheit wenden Sie sich bitte an:
If you have any questions regarding product safety, please contact:

Walter de Gruyter GmbH
Genthiner Straße 13
10785 Berlin
productsafety@degruyterbrill.com